Suborder	Superfamily	Family	Subfamily	Genus	Common name
Simiae	Ceboidea (New World monkeys, platyrrhine monkeys)	Cebidae	Cebinae	*Pithecia* *Chiropotes* *Cacajao* *Alouatta* *Saimiri* *Cebus* *Ateles* *Lagothrix*	Saki Saki Uakari Howler monkey Squirrel monkey Capuchin Spider monkey Woolly monkey
	Cercopithecoidea (Old World monkeys, catarrhine monkeys)	Cercopithecidae	Cercopithecinae	*Macaca* *Cynopithecus* *Papio* *Theropithecus* *Cercocebus* *Cercopithecus* *Erythrocebus*	Macaque Black ape Baboon, drill, mandrill Gelada Mangabey Guenon Patas monkey (hussar monkey, red monkey)
			Colobinae	*Presbytis* *Pygathrix* *Rhinopithecus* *Simias* *Nasalis* *Colobus*	Langur, leaf-monkey Douc Snub-nosed monkey Pig-tailed langur (Mentawi Islands langur) Proboscis monkey Guereza
	Hominoidea (apes and man)	Hylobatidae (lesser apes)		*Hylobates* *Symphalangus*	Gibbon Siamang
		Pongidae (great apes)	Ponginae	*Pongo* *Pan* *Gorilla*	Orangutan Chimpanzee Gorilla
		Hominidae		*Homo*	Man

NOTE: Names in parentheses are synonyms for the names they immediately follow. Names separated by commas, but not in parentheses, are not synonyms.

Contributors to This Volume

IRWIN S. BERNSTEIN

RUSSELL L. DeVALOIS

R. A. HINDE

GERALD H. JACOBS

WILLIAM C. STEBBINS

Behavior of Nonhuman Primates
MODERN RESEARCH TRENDS

Volume 3

Behavior of Nonhuman Primates

MODERN RESEARCH TRENDS

EDITED BY

ALLAN M. SCHRIER

PRIMATE BEHAVIOR LABORATORY
WALTER S. HUNTER LABORATORY
OF PSYCHOLOGY
BROWN UNIVERSITY
PROVIDENCE, RHODE ISLAND

FRED STOLLNITZ

DEPARTMENT OF PSYCHOLOGY
MORRILL HALL
CORNELL UNIVERSITY
ITHACA, NEW YORK

Volume 3

1971

ACADEMIC PRESS New York and London

ACADEMIC PRESS, INC.
111 Fifth Avenue, New York, New York 10003

United Kingdom Edition published by
ACADEMIC PRESS, INC. (LONDON) LTD.
Berkeley Square House, London W1X 6BA

LIBRARY OF CONGRESS CATALOG CARD NUMBER: 65-18435

PRINTED IN THE UNITED STATES OF AMERICA

We dedicate Volumes 3 and 4 of this series to Harry F. Harlow. His contributions to the field have been many; they have had a profound influence on us all, and without doubt will continue to be influential for many years to come.

Contents

Chapter 1

DEVELOPMENT OF SOCIAL BEHAVIOR

By R. A. Hinde

Chapter 2

ACTIVITY PROFILES OF PRIMATE GROUPS

By Irwin S. Bernstein

Contents

Chapter 3

VISION

By Russell L. DeValois and Gerald H. Jacobs

Chapter 4

HEARING

By William C. Stebbins

List of Contributors

Numbers in parentheses indicate the pages on which the authors' contributions begin.

IRWIN S. BERNSTEIN, Yerkes Regional Primate Research Center, Emory University, Atlanta, Georgia (69)

RUSSELL L. DeVALOIS, Department of Psychology, University of California, Berkeley, California (107)

R. H. HINDE, Medical Research Council Unit on the Development and Integration of Behaviour, Madingley, Cambridge, England (1)

GERALD H. JACOBS, Department of Psychology, University of California, Santa Barbara, California (107)

WILLIAM C. STEBBINS, Kresge Hearing Research Institute and Departments of Otorhinolaryngology and Psychology, University of Michigan, Ann Arbor, Michigan (159)

Preface

Volumes I and II of this series appeared in 1965. At that time, we did not have a continuing series of such volumes in mind. The original preface stated that research on nonhuman primates was mushrooming, but that there were no books for researchers and students to turn to for summaries and integration of that research. We felt that such a source was needed, and the fact that the volumes were quite well received suggests that many people agreed with us.

A great deal of research on behavior of nonhuman primates continues to be published in various places, but there are still few single sources devoted exclusively to summarizing the research and thinking in the area. Hence, the additional volumes. Our main aim is to provide chapter authors with an opportunity to describe and integrate work done by themselves and others in their special area of research. We prefer that each chapter give at least as much emphasis to the integration of data as to the presentation of new data. Not all authors can meet that goal, but to the extent that they do, we measure the success of the volumes.

As was the case when Volumes 1 and 2 were published, the nomenclature used to identify the species of subjects poses a problem. Use of vernacular names involves a variety of difficulties, but they are in such widespread use that researchers are more familiar with them than with the scientific names. Therefore, we decided to continue the procedure of the earlier volumes of using vernacular names and giving the scientific name, insofar as it could be ascertained, the first time a species is mentioned in a chapter. We refer the reader to the Appendix to Volumes I and II for a further discussion of nomenclature. One important change from those volumes is that our reference for scientific names is now J. R. Napier and P. H. Napier (*A handbook of living primates.* New York and London: Academic Press, 1967). Although many of the names they have adopted are still being debated by taxonomists (as would be true of any such attempt at a comprehensive list of scientific names), Napier and Napier is the most recent and conveniently obtainable comprehensive source of information on primate taxonomy. Some problems have arisen as a result of Napier and Napier's splitting groups of similar animals into different species rather than leaving them as subspecies. This was especially true with the aethiops group of guenons (*Cercopithecus* spp.). Evidently, what most field and laboratory researchers have called interchangeably "green monkeys" or "vervets" and have designated as *Cercopithecus aethiops* may be subdivided into *C. aethiops, C. pygerythrus,* and *C. sabaeus.* Furthermore, Napier and Napier refer to these, respectively, by the vernacular names "grivet," "vervet," and "green monkey," which have also largely been used interchangeably. Because it is usually impossible to determine which of the species was used on the basis of information provided in the original journal articles, we have referred to all animals in the aethiops group as "vervets" or "green monkeys" though we have retained the scientific name used in the original article. Two changes in scientific names that are now widely accepted may be especially noticeable to readers. The cynomolgus (or crab-eating or Java) monkey is now designated *Macaca fascicularis*, rather than *M. irus*, and the stumptailed monkey is now designated *M. arctoides*, rather than *M. speciosa*.

We thank Dorothy Brown and Gordon W. Wilcox for helpful comments on some of the chapters and Mrs. Kathryn M. Huntington for help above and beyond what might ordinarily be expected of a secretary.

Contents of Previous and Future Volumes

Behavior of Nonhuman Primates
MODERN RESEARCH TRENDS

Volume 3

Chapter 1

Development of Social Behavior[1]

R. A. Hinde

Medical Research Council Unit on the Development and Integration of Behaviour, Madingley, Cambridge, England

I. INTRODUCTION

The infant primate depends on its mother for survival. In the course of evolution, natural selection has molded the physical and behavioral characteristics of both mother and infant so that they interact in a manner which, under most circumstances, enhances the infant's chances of survival. In turn, mother and infant usually form part of a group, and under natural con-

[1] This review was written while my work was supported by the Royal Society and the Medical Research Council.

ditions may depend on their group companions for survival. Both mother and infant interact with these group companions in diverse ways that affect the behavioral development of the infant. Thus the infant monkey forms part of a social nexus, changes in many parts of which may affect it directly or indirectly.

The pioneering work of Harlow and Harlow and their collaborators at Wisconsin, reviewed in Volume II of this series (1965), has shown that the behavioral development of rhesus monkeys (*Macaca mulatta*) may be grossly distorted if they are reared without companions. These workers have also described the relationships formed when the infant is allowed access to different categories of social companions, and specified many of the factors that control the interactions involved. The present chapter reviews work that extends these studies in three ways. First, current knowledge of the social development of rhesus infants is compared with data on other species. Although detailed studies are still scarce, it is rapidly becoming clear that interspecies differences in the social influences affecting young primates may be considerable, and caution is necessary before generalizations are made. The differences are, of course, greater, the wider the phyletic spectrum considered. Within the monkeys and apes (Anthropoidea) they are more of degree than of kind: an influence on infant development which is inconspicuous in one species may be noticeable in another and of great importance in a third. The prosimians (Prosimii), so far as the scanty available information indicates, are both diverse among themselves and more different from the monkeys and apes than the latter are from each other. The young of the tree shrew (*Tupaia glis belangeri*), for instance, are left in a nest and visited by the mother only once every 48 hours (Martin, 1968): the social forces influencing their early development must be of a quite different nature from those affecting, for instance, the troop-living langur monkey (*Presbytis*). Studies of the variations on a theme to be found among the monkeys and apes are mutually revealing, and this chapter is therefore confined to them. Even within those groups, this chapter is not intended as a comprehensive review, but rather to indicate the nature and extent of the diversity to be found, and the trends in current research.

Second, attention is focused more on the nature of the interactions between the infant and its various social companions, rather than on the long-term consequences of those interactions: each type of social companion may interact with the infant in a number of ways, and in each case, both infant and social companion determine the nature of the interaction.

Third, the complexity of the social environment will be stressed. Each type of interaction with each type of social companion may be affected by both the physical and the social environment in which it occurs. Furthermore, the interactions of social companions with each other may affect each of their relations with the infant.

Data on these issues come from four main sources:

1. HAND-REARING OF INDIVIDUAL INFANTS

This can provide intimate information on infant development (e.g., Bolwig, 1963; Rowell, 1965; Kellogg, 1968). Such studies are, however, hard to evaluate, since the environmental situation cannot easily be specified or compared with the natural one.

2. CONTROLLED LABORATORY STUDIES

These permit accurate study of development under conditions in which the physical and social environment is carefully controlled. Such studies have proved very productive. In particular, the effect of adding single factors at a time (mother, peers, etc.) can be assessed (e.g., Harlow & Harlow, 1965). They have two disadvantages. First, the environment under which the animals mature is usually impoverished, both physically and socially: the relationships are often thus quite different from those that would be established under conditions more normal for the species. Second, the study of one social factor at a time distracts attention from the interactions between them.

3. STUDY OF CAPTIVE GROUPS

Here the animals live in an environment of moderate complexity, and a moderate degree of experimental control and moderately precise quantification is possible (see Chapter 2 by Bernstein). For instance, the data from Madingley, Cambridge, quoted in subsequent sections, were obtained from small groups of rhesus monkeys, each consisting of a male, two to four females and their young, living in an outside cage 18 x 8 x 8 ft connecting with an inside room 6.0 x 4.5 x 7.5 ft. Such conditions have both the advantages and the disadvantages of a compromise situation. They do, however, permit an alternative approach to the problem of social factors in development in that the effect of removing, rather than adding, single factors at a time may be studied.

4. FIELD STUDIES

These provide essential information on interspecies diversity. They also permit an assessment of the complexity of the social environment to which the infant is adapted (in an evolutionary sense) (e.g., Hamburg, 1969). Experimental control is, however, usually impossible, and precise quantification at best laborious (see Plutchnik, 1964). Taming is sometimes possible, especially if provisioning techniques are used: the advantages of these must be weighed against their possible social and ecological consequences.

In some cases, studies by one worker of the same species in captive groups and under natural conditions have enabled the differences to be assessed (e.g. the hamadryas baboon, *Papio hamadryas,* Kummer & Kurt, 1965; the patas monkey, *Erythrocebus patas,* Hall & Mayer, 1966; and the olive baboon, *P. anubis,*

Rowell, 1967): as far as the problems discussed in this chapter are concerned, they appear not to be very considerable. An important exception, however, may lie in the amount of restrictive behavior shown by mothers, which Rowell found to be much greater in caged groups than in the field. Maternal rejection of the infants' attempts to gain the nipple may also be enhanced by caged conditions (see also Baldwin, 1969, for a comparison of field data on the squirrel monkey (*Saimiri sciureus*) with that obtained by other workers in a laboratory setting).

No one of these methods is superior to the others: all can contribute, and the complex problems of social development will not be solved without the use not only of all these methods, but also of many others. One of the aims of this chapter is to show how the different methods are beginning to illuminate each other.

To facilitate this process in the future, there is a clear need for much greater standardization of recording and scoring techniques than occurs at present. In particular, two points may be mentioned. First, compound scores (e.g., "positive behavior of mother to infant"), even if adequately defined, depend too heavily on value judgments by the observer, are unnecessarily motivationally heterogeneous, and are almost impossible to use for comparative purposes. Second, quantitative data from field studies are becoming increasingly important. As field studies of primates emerge from the descriptive phase and become oriented toward problems, it is increasingly important that they should yield quantitative data, and that those data should be collected efficiently. For instance, for many purposes data on animals not engaged in specified activities are as essential as data on animals which are: it may be important to know not only how many animals of each age and sex class were seen grooming and being groomed, but also the relative probability of members of each class being so engaged, the relative proportion of the grooming activity of each class directed towards each class, and the relative proportion that would be so directed if all classes were equally abundant (e.g., Chalmers, 1967; van Lawick-Goodall, 1968). At the same time, of course, it is necessary to avoid bogus quantification, and to allow for differential visibility of the age and sex classes when engaged in different activities (see also Crook, 1970).

One other general point requires emphasis. The analysis of behavior repeatedly demands that we classify: just because we must name items of behavior, the external factors which influence it, and the internal processes supposedly involved in its determination, we must first classify them. Too often, however, the categories we use either falsify reality, or turn out to be irrelevant to the problem in hand. Nowhere is this more true than in the study of development (Hinde, in press). For example, we classify processes internal to the organism that are supposed to determine behavior, such as growth, differentiation, maturation, learning, but as soon as we do so, we are in danger both of considering each as causally homogeneous, distinct from the others, and of forgetting their interactions. Again, in considering social influences on development, we naturally classify them in terms of the units we manipulate: mother, peers, and so on. As we shall see, however, such a classification does not necessarily correspond with the effects on the infant (milk, contact comfort, play stimuli, etc.). Another type of classification of immediate concern here is that involved in the division of the continuous stream of development into a series of stages. This is discussed in the next section.

II. STAGES OF DEVELOPMENT

One difficulty in comparing the social development of different species is that of finding a proper datum, since the relations of both physical and social development to chronological age vary widely. Some authors have found it useful to distinguish "stages" in terms of physical development or social independence or both. Some examples, with the characteristics considerably abbreviated from those used by the original authors, are shown in Table I.

Description of development in terms of stages is clearly of most value to the field worker, for it enables animals to be assigned to age groups by their appearance and behavior. However, it has certain limitations and dangers. First and most obviously, the stages inevitably intergrade, so that the distinctions between them are never clear cut. Although this is always recognized by those who first describe them, the categories soon become reified and distract attention from the continuity of development. Second, and in part arising from the first, the ages given are necessarily approximate (a difficulty not helped by ambiguities which arise in labeling, as for instance between n-month-old infants and nth month infants). Third, the stages are often based on multiple criteria (e.g., both physical and social development) that may become dissociated if the conditions of development are changed. Kummer's (1968b) stages for the hamadryas baboon are a notable exception here, since they are based solely on physical criteria. Fourth, there may be considerable differences between the stages used by different authors for the same species. This is shown in Table I by the differences between the stages used for chimpanzees by Reynolds and Reynolds (1965) and, on the basis of more extensive data, by van Lawick-Goodall (1968; Goodall, 1965). Fifth, the stages are of little use for cross-species comparisons. Thus those used for langur and baboon by Jay and DeVore (Table I) correspond in name and approximately in age, but, from the descriptions given, in the earlier stages the species could differ considerably in independence from the mother. Again, an Infant 2 baboon is clearly a much more independent animal than an Infant 2 chimpanzee as described either by Goodall or by the Reynolds. Finally, a stage may be descriptively valid, and yet the criteria by which it is defined be of no or of minor importance for subsequent development (Sackett, personal communication).

TABLE I. Some Examples of the "Stages" Used in Descriptions

Howler *(Alouatta)* (Carpenter, 1965)	Baboon *(Papio)* (DeVore, 1963; Hall & DeVore, 1965)	Baboon *(P. hamadryas)* (Kummer, 1968b)	Vervet *(Cercopithecus* sp.*)* (Gartlan & Brain, 1968)
Infant 1. 0 to 5–6 months. Almost constantly on mother's belly. Greyish-brown.	*Newborn.* 0 to 1 month. On mother almost continuously.	*Black males and females.* 0 to ½ year. Hair completely or partially black.	*Infant 1.* 0 to 4–5 months. Physically small. Dark in colour, face and ears pink. Leaves mother rarely. Not weaned fully.
Infant 2. 5–6 to 10–12 months. Closely associated with mother. Brownish-black.	*Infant 1.* 1 to 4 months. Occasionally moves away from mother.	*One-year-olds.* ½ to 1½ years. Sitting height *c.* 30 cm. Brown hair.	*Infant 2.* 4–5 to 18 months. Physically immature. Peer oriented. Found in play groups. Returns rarely to mother. Weaned.
Infant 3. 10-12 to 18-20 months. Occasionally travels alone. Black.	*Transition.* 4 to 6 months. Colour changes. Much play. Eats solid food. Goes up to 20 yards from mother.	*Two-year-olds.* 1½ to 2½ years. Sitting height *c.* 40 cm. Brown hair.	
		Three-year-old males. 2½ to 3½ years. Sitting height *c.* 47 cm. No signs of mantle, but in some cases longer hairs at the sides of the head.	*Juvenile female.* 18 months to 2 years. Not fully grown. Alarm barks given readily.
Juvenile 1. 20 to 30 months. Weaning. Relatively independent. Often with play group. Black.	*Infant 2.* 6 months to 1 year. Increasing independence. Much time with peers.		*Juvenile male.* 18 months to 4 years. Physically not fully grown. Scrotum small and purple. Voice rather high pitched. Alarm barks given readily.
Juvenile 2. 30 to 40 months. Usually with peers, only occasionally with mother. Black with reddish mantle.	*Weaning.* 11 to 15 months.	*Three-year-old females.* 2½ to 3½ years. Sitting height *c.* 45 cm. Nipples always button-like.	
Juvenile 3. 40 to 50 months. Entirely independent. Black with distinctly red mantle.	*Young juvenile.* 2nd year. Spends most of day with peers. Flees to male, not mother.	*Sub-adult males.* Range from sitting height 56 cm, first signs of mantle (3½ to 4 years), to *c.* 60 cm, mantle fully developed (5–7 years).	*Sub-adult female.* 3rd year. Physically mature and fully grown. Nipples not elongated.
	Older juvenile. 3rd and 4th years.	*Sub-adult females.* 3½ to 5 years. Sitting height *c.* 50 cm. Nipples usually short, not longer than wide.	

of the Development of Behavior of Nonhuman Primates

Langur *(Presbytis entellus)* (Jay, 1963; abbrev. in Jay, 1965)	Chimpanzee *(Pan)* (Reynolds & Reynolds, 1965)	Chimpanzee *(Pan)* (van Lawick-Goodall, 1968)	Gorilla *(Gorilla)* (Schaller, 1963, 1965)
Newborn. 0 to 1 month. Goes a few feet from mother only.	*Infant 1.* 0 to 6 months. Clings continuously.	*Infant 1.* 0 to 6 months. Almost continuously on mother.	*Infant.* 0 to 3 years. Carried by or rides on female for prolonged periods. Weighs $<c.$ 60 lb.
Infant 1. 1 to 3 months. Wanders up to 10 ft from mother.	*Infant 2.* 6 months to 2 years. Tiny. Active at short distances from mother.	*Infant 2.* (Early) 6 months to 2 years. Takes some solid food. Rarely out of sight of mother. Conspicuous rump tuft.	*Juvenile.* 3 to 6 years. Larger than an infant and smaller than an adult. Lacks prolonged close contact with female. 60–120 lb.
Transition. 3 to 5 months. Colour changes. Spends less time with mother, more with peers.	*Juvenile 1.* 2 to 4 years. One quarter adult size. Independent for hours at a time. Carried by mother when moving.	*Infant 3.* 2 to 3½–4 years. Diet like adult. Still protected by mother, but gradually rides on her back less often.	*Sub-adult and adult.* 6+ years. Advanced beyond juvenile stage.
Infant 2. 3–5 months to 1 year. Plays with peers for several hours each day. Eats by itself.	*Juvenile 2.* 4 to 6 years. One third adult size. Occasionally carried by mother.	*Juvenile.* 3½–4 to 7 years. Independent of mother for feeding, transport and sleeping. Rump tuft gradually becomes less conspicuous.	
Weaning. 11 to 15 months.	*Adolescent.* 6 to 8 years. One half adult size. May follow mother, but usually on own.	*Adolescent.* 7–9 to 10–12 years (female), 7 to 13 years (male). Rump tuft disappearing.	
Young juvenile. 2nd year. Spends most of time with peers.			

The division of development into stages, though useful for some purposes, can thus bring difficulties for comparative and experimental studies. A firmer basis for such purposes would be provided by more extensive data on the ages at which young animals reach various criteria of physical and social development under defined conditions. Some potentially useful examples are shown in Table II. The figures given refer to the earliest age at which the author cited saw the criteria reached, except where the median and range for a number of animals are given. They were in some cases extracted from reports of field studies in which, presumably, the author was not primarily interested in obtaining data of this sort, and are of widely varying reliability. They illustrate (a) the need for precise definition of criteria of development that would be valid for cross-species comparisons, (b) the present scarcity of data, and (c) the unreliability of some of the published information. However, the data are already sufficient to indicate the considerable diversity, even among monkeys, in the ages at which the criteria are attained. Even though some of the apparent variation is due to inadequacies in the data available at present, further accumulation of information of this sort is at least capable of providing a chronological framework for discussions of behavioral development.

III. THE CONSEQUENCES OF SOCIAL DEPRIVATION

That the social conditions of early rearing have a marked effect on behavioral development is now well established. Much of the work on this subject was reviewed by Harlow and Harlow (1965), and only the principal findings are sketched here (see also Sackett, 1968a, b):

1. A comparison between (a) rhesus monkeys captured in the field and (b) animals born in the laboratory, removed from their mothers within a day or two of birth, hand fed for 15–20 days, and subsequently raised in wire cages with visual and auditory but no tactile contact with other monkeys, revealed considerable differences between the two groups in social behavior, and in responses to novel environments and to alien species (Mason, 1960, 1961a, b, 1962).

2. Rhesus monkeys brought up under conditions of severe social isolation developed into neurotic and asocial adults. The abnormalities included orality, self-clutching, deficiencies in sexual and maternal behavior, with social indifference, hyperaggression, or both (Harlow & Harlow, 1965, 1969). They were not correlated with any marked intellectual deficits (Harlow & Harlow, 1969; Harlow, Schiltz, & Harlow, 1969), but were extremely persistent (Mitchell, Raymond, Ruppenthal, & Harlow, 1966; Mitchell, 1968b; Sackett, 1968a, b; see also Evans, 1967, for data on pigtailed monkeys, *Macaca nemestrina*). The severity of the effects depended on the period of isolation. Total isolation for 3 months produced depression on removal from isolation, but

after recovery from this, there were no marked permanent abnormalities in social behavior (Griffin & Harlow, 1966). Isolation for 6 months produced considerable permanent effects, and isolation for 1 year destroyed all social abilities (Harlow & Harlow, 1969; Sackett, 1968a, b).

In part because a smaller group of rhesus monkeys reared in social isolation at the Puerto Rico laboratory have been said to show less severe effects (Meier, 1965), it has been suggested that the effects of isolation are due to the trauma of the testing procedures, the previously isolated animals lacking adaptation to the novel situations encountered (Fuller, 1967; Jensen & Bobbitt, 1967). However, this is unlikely on several counts. First, the conditions under which the young animals were reared differed considerably between the two laboratories: the Puerto Rican animals were permitted much more visual and auditory contact with each other than the Wisconsin ones. Second, precise comparison between the two groups of animals has not been possible, and further study of the Puerto Rican animals revealed a considerable number of abnormalities (Missakian, 1969). Finally, the importance of trauma of testing procedures in producing the abnormalities in the Wisconsin animals is rendered improbable by the findings that individual adaptation to the testing environment is insufficient to overcome the effects of isolation, while isolation preceded by experience with peers and adaption does not produce marked abnormalities (Clark, cited by Sackett, 1968a; Sackett, 1968a).

3. Rhesus infants raised by their own mothers but denied access to peers showed deficiencies in affectionate behavior and hyperaggressiveness. The effects were present if access to peers was denied for 4 months, and marked after 8 months. However, the effects of 8 months' peer deprivation were less severe than those of 6 months' rearing in the absence of both mothers and peers (Alexander, cited by Harlow & Harlow, 1969). Infants reared for 1 year in large cages without peers and subsequently placed in a group showed social interactions that were qualitatively similar to but less frequent than those of group-reared infants (Spencer-Booth, 1969).

4. Rhesus infants raised without mothers but in the presence of peers quickly developed physical attachments to each other. Play occurred earlier in groups of four or six than in groups of two, but later than in normally reared animals. The long-term social adjustments of such animals appear not to be markedly deficient, peer–peer interactions apparently compensating for the lack of mothering. Animals reared with the same individual peer throughout formed strong partner-ties but less carryover to peers encountered subsequently, than infants reared with frequently changing peers, which showed more play and less aggression both within the group and towards outsiders (Chemore, cited by Harlow & Harlow, 1969).

Rhesus monkeys raised with visual and auditory but no tactile contact with peers may be rehabilitated by subsequent experience with peers. Such

TABLE II. Summary of Data

Species	Authority	Type of study	Releases grip for tactile exploration	Coordinated reaching	Eye-hand-mouth coordination
Marmoset *(Callithrix jacchus)*	Epple, 1967	Lab. group			
Tamarin, Pinché *(Saguinus geoffroyi, S. oedipus)*	Epple, 1967	Lab. group			
Howler *(Alouatta villosa)*	Carpenter, 1934, 1965 / Altmann, 1959	Field / Field		10 days	
Cebus *(Cebus apella)*	Nolte & Dücker, 1959	Lab. $N = 1$	2nd week	2-3 weeks	16 days
Squirrel Monkey *(Saimiri sciureus)*	Hopf, 1967 / Vandenbergh, 1966 / Rosenblum, 1968 / Baldwin, 1969	Lab. / Lab. group $N = 1$ / Lab. $N = 6$ / Field	2nd week	*c.* 2 weeks / 1st week / *c.* 2 weeks / 3rd week	
Rhesus *(Macaca mulatta)*	Hinde *et al.,* 1964 / Hinde & Spencer-Booth. 1967a / Southwick *et al.,* 1965	Lab. groups / Field $N = 1$	5(3-6) days	9(6-10) days	18(16-24) days
Stumptail *(Macaca arctoides)*	Bertrand, 1969 and personal communication	Lab. & zoo $N = 1–5$	Day 3-6	*c.* 2 weeks	*c.* 16 days
Mangabey *(Cercocebus albigena)*	Chalmers, 1967 / Chalmers, personal communication	Field $N = 1–2$ / Lab. group $N = 4$		36(34-40) days	2½-3½ weeks / 44(34-51) days
Baboon *(Papio ursinus, P. anubis)*	DeVore, 1963 / Hall & DeVore, 1965 / Rowell *et al.,* 1968 / Rowell, personal communication	Field / Field / Lab. colony $N = c.$ 6	3 days	3 days	8 days
Vervet *(Cercopithecus aethiops)*	Chalmers, personal communication	Lab. group $N = 5$		20(8-44) days	39(26-54) days
Vervet *(Cercopithecus sabaeus)*	Schlott, 1956 / Moog. 1957.	Zoo	15 days		
Sykes' monkey *(Cercopithecus mitis)*	Chalmers, personal communication	Lab. group $N = 2$		29, 34 days	34, 44 days
Talapoin *(Cercopithecus talapoin)*	Hill, 1966	Lab. $N = 1$	2 days	4 days	4 days
Allen's Swamp Monkey *(Cercopithecus nigroviridis)*	Pournelle, 1962				
Patas Monkey *(Erythrocebus patas)*	Goswell & Gartlan, 1965	Lab. $N = 1$	Day 2	Day 5	Day 5
Hanuman Langur *(Presbytis entellus)*	Jay, 1963, 1965 / Sugiyama, 1965a / Yoshiba, 1968	Field / Field / Field		7 days	17 days
Chimpanzee *(Pan troglodytes)*	Van Lawick-Goodall, 1968 / Mason, 1965b	Field / Lab.		6-10 weeks	
Gorilla *(Gorilla gorilla)*	Schaller, 1963	Field, zoo		7-11 weeks	

The figures in the body of the table refer to first ages at which criteria were seen to be reached.

on Behavioral Development

Eating solid food	Off mother	More than 2 ft from mother	Steady walking	Competent climbing off mother	Sexual or pseudo-sexual behavior	Weaning
26-35 days	17-21 days (all adults)					60 days
40 days	23 days (all adults)					
2-3 weeks 31 days	15 days (all adults)		c. 1 month	15-20 days		1½-2 years
6 weeks 4 days	c. 5 weeks			6-7 weeks		
	c. 3 weeks				4-5 months	c. 9 months
7 weeks	14 days			2 weeks		
3rd month	c. 3 weeks					
	5 weeks		6 weeks	6 weeks	8-10 weeks	8-11 months
21(13-31) days	9(4-15) days	1-2 weeks	2-3 weeks	3-4 weeks	Mounting: ♂ 12th-46th week Presenting: ♂ 25th-30th week ♀ 13th-18th week	Variable 3-13 months
	4 days	7 days				By 12 months
27, 38 days	c. 15 days	20-21 days	3-4 weeks		Mounting: ♂ 9-11 weeks Presenting: ♂ 7-12 weeks ♀ 7 weeks	Variable c. 1 year
	2-9 days 4-11 days	6 weeks	3-4 weeks	c. 6 weeks		
47(23-51) days	14(9-36) days		44(35-51) days	22(12-36) days		
5-6 months	c. 4 weeks					11-15 months
3-4 weeks	<1 week		c. 21 days	c. 20-35 days		
26(18-48) days	6(3-8) days		31(26-49) days	15(8-29) days	Mounting 81, 89 days	
30 days	21 days			27-30 days		60 days
29, 54 days	4, 14 days		34, 34 days	14, 26 days		
16 days	6 days		c. 8 days	17 days	54 days	50 days
33 days	14 days					75 days
? Day 12	Day 8					
By 3 months	1-2 weeks (all adults)	c. 2 weeks	c. 3 weeks			11-15 months
27th day	9 days (all adults)	c. 4 weeks	8th week	c. 2 months	c. 10 months	Up to 20 months
						11-15 months
c. 5 months c. 6 months	14-22 weeks	7-9 months	c. 12 months 4-6 months		9 months	2-5 years
2½ months	c. 3 months		5-6 months	c. 6 months		

Ranges are indicated in brackets and medians precede the brackets. For reservations, see text.

experience is most beneficial if animals of similar social competence are tested together (Sackett, 1968a).

5. Infants reared on surrogate mothers and allowed access to peers showed near-normal social behavior though, during their first half-year, they tended to be slightly retarded in comparison with mother-raised infants (Rosenblum & Hansen, cited by Harlow & Harlow, 1969; Hansen, 1966).

6. Whereas social experience after 3 months of age seems to be most effective in producing normal development, learning during the first month of life can produce preferences for specific social stimuli that persist into adult life. This effect can, however, be overridden by social experience during the rest of the first year of life (Sackett, 1968a).

These experiments show that fairly severe social deprivation during the first year of life can have a profound influence on later development, that the effects may be more severe at some ages than at others, and that the effect of a given type of social experience during one age span may depend on experience at others. So far, such data are largely limited to macaques, though marked differences between wild-reared and isolation-reared animals are known to occur also in other species (e.g., chimpanzees, Randolph & Mason, 1969). We shall consider other consequences of less profound changes in the social environment subsequently. For the main part, however, the remainder of this chapter will be concerned with the nature of the interactions of the infant primate with various categories of social companion.

IV. INTERACTIONS WITH SOCIAL COMPANIONS

A. The Mother

1. Age Changes in the Mother-Infant Relationship in Group-Living Rhesus Monkeys

The data on the development of the mother–infant relationship in rhesus macaques given by Hansen (1966) and Harlow and Harlow (1965) were obtained from animals living in small laboratory cages with restricted access to social companions. This species has also been studied in natural troops by Southwick, Beg, and Siddiqi (1965), Korford (1965), Kaufmann (1966) and others, and in small captive groups by Hinde, Rowell, and Spencer-Booth (1964) and Hinde and Spencer-Booth (1967a, 1968). The age changes in certain aspects of mother–infant interaction in the latter situation are shown in Fig. 1. These data were obtained from eight infants studied over their first 2.5 years of life and a further eight studied for their first 24 to 30 weeks. All watches were made between 0900–1300 hours, each infant being watched for 6 hours per week, a fortnight, or a month, according to age. The data were collected on check sheets by 30-second periods.

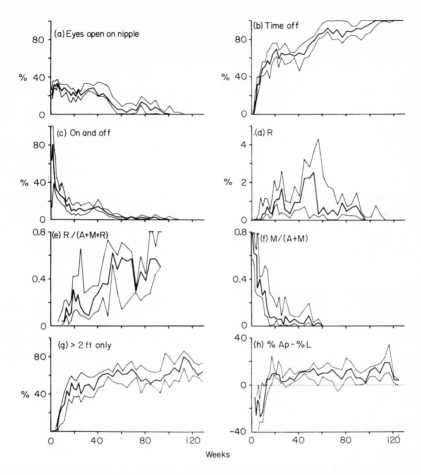

FIG. 1. Age changes in mother–infant interaction. (a) *Eyes open on nipple.* Number of half minutes in which awake on nipple. (b) *Time off.* Number of half minutes in which recorded off mother. (c) *On and off.* Number of half minutes in which recorded both on and off mother as a percentage of the number in which they were off at all. (d) *R.* Number of times rejected by their mothers. (e) $R/A + M + R.$ Ratio of number of times infants were rejected (R) to total number of times infants attempted to gain nipple ($A + R$) or were picked up by their mothers (M). (f) $M/A + M.$ Ratio of number of times infants were picked up by their mothers *(M)* on mothers' initiative, to total number of times they were accepted (A) or picked up by them (M). (g) *> 2 feet only.* Number of half minutes which infants spent wholly more than 2 feet from their mothers as a percentage of the number in which they were off at all. (h) $\%Ap - \%L$. Difference between percentage of approaches (from more than 2 feet to less than 2 feet) which were due to movement by infants, and percentage of leavings (from less than 2 feet to more than 2 feet) which were due to infants.

In (a), (b), and (d), data are expressed as a percentage of half minutes watched. In each case the thick line represents the median and the thinner lines the limits of the interquartile range. (After Hinde & Spencer-Booth, 1968)

In summary, the time spent awake and on the nipple (Fig. 1a) decreased little over the first 40 weeks, but then fell gradually to zero: the infant was, of course, actually sucking for only a small proportion of the time that it was attached to the nipple. The initial rapid increase in time off the mother (Fig. 1b) was due primarily to a decrease in time spent asleep from about 75% in the first week to about 20% in the tenth. The time off the mother continued to increase rapidly until about Week 20. It then remained fairly steady, to increase again to nearly 100% at the end of the first year. This was accompanied by an increase in the length of bouts off the mother (i.e., a decrease in the number of 30-second periods in which infant was both on and off mother per the number of 30-second periods in which it was off at all; *on and off*, Fig. 1c).

The frequency of maternal rejections (R, Fig. 1d) was low in the first 2 months, reached a peak in weeks 15 to 25, and a second one around the end of the first year, after which it fluctuated at a low level. The relative frequency of rejections (i.e., number of rejections per number of times infant gained or attempted to gain the nipple; $R/A + M + R$, Fig. 1e) changed in a somewhat similar fashion, except that the high level reached at the end of the first year was maintained. Maternal initiative in nipple contacts decreased rapidly during the early weeks ($M/A + M$, Fig. 1f).

The proportion of time the infant spent out of the mother's immediate reach (i.e., number of 30-second periods it spent wholly more than 2 feet from the mother as a percentage of those in which it was off her at all; more than 2 feet only, Fig. 1g) increased to about 50% after about 20 weeks, and thereafter increased only slowly. The role of the infant in maintaining proximity to the mother is expressed here as the difference between the proportion of approaches (distance between mother and infant decreases from more than 2 feet to less than 2 feet), which were due to the infant and the proportion of leavings (distance increases from less than 2 feet to more than 2 feet), which were due to the infant (percentages of approaches minus percentages of leavings ($\%Ap - \%L$), see Section IV, A, 3). This was initially negative (i.e., mother was primarily responsible for maintaining proximity), but this changed gradually and after about Week 20 the position was reversed.

The increase in rejections by the mother up to Week 20 was associated with a decrease in the frequency with which the mothers restrained their infants when exploring and climbing, an increase in the frequency with which they hit them, and an increase in the percentage of infants showing tantrums. The amount of maternal grooming, however, remained fairly steady (median about 3% of 30-second periods) until the end of the 2.5-year period.

Hansen (1966) and Harlow and Harlow (1965) discussed the mother–infant relationship in terms of a number of stages: they recognized four stages in the infant's relationship to its mother (reflex, comfort and attachment, security, and separation), and three in the mother's relationship to the infant (attachment and protection, ambivalence, and separation). These authors were, however, at pains to point out that the transition between these stages is gradual, and that they overlap extensively. The data in Fig. 1 give only limited support to a division of development into a series of stages, for most of the measures changed gradually. The only age at which there might appear to be a marked change in the nature of the relationship is about 20 weeks, where the time off the mother reaches a temporary plateau, and the roles of mother and infant in maintaining proximity are reversed. However, the individual records showed considerable variability on this issue, and it is in any case markedly influenced by the physical and social environment (see Sections IV, A, 7 and IV, B). On balance, the division of development into stages distracts attention from its essential continuity, and the mother–infant relationship is better regarded as involving ambivalence on both sides almost from the start, with the balance gradually changing as the infant grows (see also Jensen, Bobbitt, & Gordon, in press).

Unfortunately, differences in conditions and techniques render more than qualitative comparison with the Wisconsin data (e.g., Harlow & Harlow, 1965; Hansen, 1966; Seay, 1966) impossible. In particular, in most of the latter studies the infants could leave their mothers through a door too small for the mother to follow. The mothers could thus not retrieve or approach their infants. Instead two expressive movements, the grimace (similar to the fear grin; see Hinde & Rowell, 1962) and the sexual present, seem to have been used to summon the infants. These were not seen in this context in the Madingley study. By and large, however, there seem to have been no major differences between the developmental trends shown in these studies.

2. MOTHER–INFANT RELATIONS IN OTHER SPECIES

Similarities in conditions and techniques are equally necessary before detailed comparisons among species are possible. We may consider first here three cases in which one observer or team of observers have studied different species under similar conditions.

Rowell, Din, and Omar (1968), studying captive baboons up to 3 to 4 months old in Uganda, used techniques similar to those used in the rhesus monkey studies at Madingley, making comparison with the latter possible. Baboons spent less time asleep than did rhesus macaques during the first fortnight, and thereafter slightly more, whereas the reverse was true for the time on the nipple with the eyes open. For most of the 3 months the baboons spent rather more time off their mothers than did the rhesus infants, but the proportion of this time spent out of arms' reach of the mother was similar. Differences in mother–infant interaction between the two species were small, though they could suggest a faster maturation rate in baboons (Rowell et al., 1968).

Data for two other macaque species living in small captive groups, the pigtailed and bonnet (*Macaca radiata*) macaques, are given by Rosenblum and Kaufman (1967) and Kaufman and Rosenblum (1969). Many of the measures are comparable to those used in the Madingley data. The general course of the mother–infant relationship in these species, and especially in the pigtailed macaque, was broadly similar to that in the rhesus. There were, however, interesting differences between them (Fig. 2). Although the amount of time

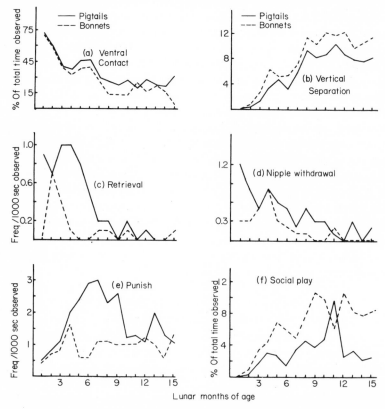

FIG. 2. Development of behavior in 8 bonnet and 9 pigtailed mother–infant pairs. (a) Median duration of ventral contact during first 15 months of life. (b) Median duration of vertical departures. (c) Relative frequency of retrievals by mothers. (d) Relative frequency of nipple withdrawal by mothers. (e) Relative frequency of punitive deterrence by mothers. (f) Mean duration of social play by infants. (After Kaufman & Rosenblum, 1969)

mother and infant spent in contact was similar, from the third month bonnet infants spent more time at a distance from their mothers than did pigtailed infants. The difference was especially marked when time spent at different levels in the pen was compared. The bonnet mothers restrained, guarded, and retrieved their infants less than did pigtailed mothers. They also withdrew the nipple and punished their infants less often. These differences seem to be related to the more placid interindividual relations among bonnet monkeys. They sit more frequently in close contact with each other, and permit other individuals greater access to their young (see Section IV, B). The nature of the relation between temperament and mother–infant interaction, and the direction of the cause-effect relationship, are, of course, open issues. More recent data by Rosenblum (1968) permit comparison of these macaque species with the more distantly related squirrel monkey.

A third comparative study (Chalmers, personal communication) used captive groups of the predominantly terrestrial vervet (*Cercopithecus aethiops*), the predominantly arboreal Sykes' monkey (*C. mitis*) and the totally arboreal mangabey (*Cercocebus albigena*). Recording methods were similar to those in the rhesus and baboon studies just cited. The time courses of visuomotor and locomotor development in these species are shown in Table II. During their first 3 months of life they spent similar amounts of time asleep. When awake, however, the mangabey and Sykes' monkey infants tended to remain in contact with their mothers more (Fig. 3a), and when off tended to remain within 2 feet of them more than did the vervets. That this was primarily due to the greater independence of the vervet infants, rather than to greater restrictiveness by the mothers of the other species, was indicated by several types of evidence: mangabey and vervet mothers were recorded holding their babies for a similar proportion of the time, and the mangabey mothers prevented attempts by their infants to leave on a rather smaller proportion of occasions than did the vervet mothers (Fig. 3b). The Sykes' monkey mothers did tend to be rather more restrictive than did the vervets, but only in the early weeks: It is unlikely that this could have accounted also for the later difference in mother-infant relations. Chalmers tentatively suggests that the greater independence of the vervet infants is related to their terrestrial habits, since the young of arboreal species may be subject to greater hazard if they move away from their mothers. An experimental approach to such interspecies differences that attempted to specify their bases more precisely would be of the greatest interest.

Data for other monkeys are not yet adequate to permit more than qualitative comparisons but, with exceptions that will become apparent later, the course of mother–infant relations seems to be generally similar in most species studied so far. The principal areas of divergence are as follows:

a. Means by which the mother holds or carries the baby. In a few species the infant is often carried in positions other than the usual ventro–ventral one. Thus the infant squirrel monkey clings to its mother's neck (Baldwin, 1969; Ploog, 1966; Rosenblum, 1968) and some guerezas (*Colobus*) are said to carry the infant in the mouth (Booth, 1957). The latter observation, if substantiated, would probably be unique among the monkeys and apes, though such behavior is not infrequent among prosimians (Sauer, 1967). In many species, especially

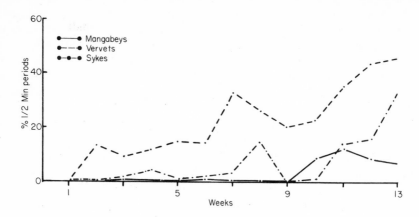

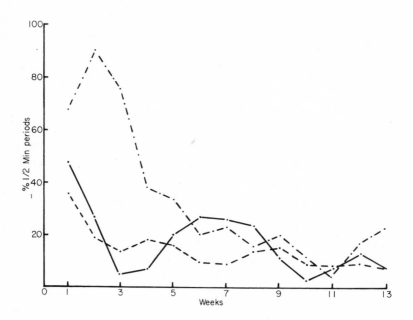

FIG. 3. Mother–infant interaction in vervet monkeys ($N = 5$), Sykes' monkeys ($N = 2$), and black mangabeys ($N = 2$). (After Chalmers, personal communication.) Top: time totally off mother as percentage of time awake. Bottom: percentage of half minutes for which mother holds.

predominantly terrestrial ones, the infant comes to ride on the mother's back rather than on her belly, but this seems to vary with the environmental conditions (e.g., it is unusual in captive rhesus groups, but apparently common in the wild) and even amongst natural groups in different parts of the species range (e.g., patas monkeys: Hall, 1965).

b. *Extent of maternal behavior shown by other social companions.* In some species the male or other troop females spend much time with the infant. This is discussed in later sections.

c. *Degree of maternal attachment and/or permissiveness.* Some examples of interspecies differences in these characters were mentioned above. They clearly have great bearing on the relations with other social companions, and will be discussed again later.

d. *The rapidity and severity of weaning.* Little is known of the factors controlling this, but published accounts suggest considerable interspecies differences. In the langur, for instance, Jay's (1963) account suggests that it is rather more traumatic than in rhesus, but in many other species it is probably much less so. The age and rapidity of weaning may depend in turn on the frequency with which the females breed. Whereas most females of most species seem able to breed yearly, female baboons in Kenya are said to become pregnant only once every 2 years (DeVore, 1963). In some species, however, the elder sibling is sometimes allowed to suckle alongside the newborn (e.g., langur, Sugiyama, 1965a; Yoshiba, 1968; baboon, Ransom & Ransom, personal communication).

Among the apes, the young are dependent on their mothers for much longer than monkeys. For instance, young chimpanzees do not first break contact with their mothers until 16–22 weeks old, continue to ride fairly frequently on their mothers' backs until well over 2 years old, and are not weaned until 3.5–4.5 years old (van Lawick-Goodall, 1969, see Table II). They may associate with their mothers frequently for many years after weaning (Fig. 4). Although detailed comparisons with monkeys are thus of doubtful value, careful study of chimpanzee mother–infant relations by van Lawick-Goodall (1967, 1968) and of gorillas by Fossey (personal communication) permit specification of some principal areas of differences for these two species.

a. The infants do not use the mother's nipple as a point of support. Chimpanzee infants can cling unsupported only very briefly for some days or weeks after birth. By contrast, Fossey reports a newly born gorilla that clung to its mother unsupported by her for up to 1.5 minutes at a time.

b. The ways in which mothers support their infants are more diverse than those used by monkeys. For instance, chimpanzee mothers may use their feet, and gorillas their elbows.

c. Maternal care is more active than in monkeys: for instance, the mother plays a considerable role in encouraging the infant to ride on her back.

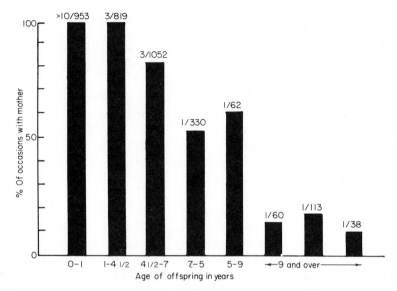

FIG. 4. Association between mother and infant in free-living chimpanzees. Number of occasions mother and infant were seen together, as percentage of total number of occasions when either or both were seen (Number of infants/Number of occasions mothers and infants were seen together shown at top of each column). (After van Lawick-Goodall, 1968)

Chimpanzee mothers may support the infant's first steps, and gorillas aid their infants in trees more than do monkeys.

d. Maternal care seems more "purposeful" (in a colloquial sense) than in monkeys. For instance, mother chimpanzees may protect their infants not only by threatening off rougher playmates, but by distracting the latter by playing with them.

e. Physical punishment of the infant by chimpanzee mothers is rarer than in monkeys, and mothers use distraction in controlling them. Whereas in many species, monkey mothers are remarkably tolerant of their infants, chimpanzee mothers seem even more so. They may, for instance, wait for their infants to leave a play group before moving off. Schaller (1953) records that gorilla mothers are likewise tolerant and gentle with their infants, but Fossey was impressed by the individual differences in maternal behavior in this species—some mothers being very permissive while not infrequently others used whacking and mock-biting.

f. Mothers sometimes share food with their infants, who may beg for it. When a mother resists an importunate infant she usually does so by preventing access to the food. Although she may push the infant away gently, she does not bite or hit it.

g. Affectionate behavior between mother and infant is conspicuous.

h. Mothers often initiate play with their infants.

i. The mother grooms the infant little during its first few weeks. [The wild-living chimpanzees studied by van Lawick-Goodall seem to differ in this respect from captive animals, and also from wild-living gorillas (Schaller, 1963)].

j. In monkeys, weaning is sometimes a traumatic business, though species and individual differences are considerable. In chimpanzees, it is usually gentle and protracted.

3. THE NATURE OF THE MEASURES OF THE MOTHER–INFANT RELATIONSHIP

Each of the measures used in Fig. 1 is necessarily influenced by the behavior of both mother and infant (see also Bell, 1968). For instance, the behavior of both partners determined the time the infants spent off their mothers, or the proportion of this time spent at a distance from her. The frequency with which the mothers rejected their infants fell off after the end of the first year, but this was in part a consequence of decreasing demand: the relative frequency $(R/A + M + R)$ stayed high. Even where a measure appears to be a reflection of the relative roles of the two partners, as for instance with the proportion of nipple contacts initiated by the mother, the apparent role of each is a consequence not only of its own motivational condition, but of the feedback it has received from the other. This interactive nature of the mother–infant relationship is usually obvious, but even where the mother's role appears to be largely passive, her behavior is in fact much influenced by that of the infant (Rumbaugh, 1965). A method for analyzing the roles of the two partners in producing changes in the relationship when only measures of interaction are available is discussed in the next section.

An additional difficulty arises in the interpretation of proximity measures. If the nature of the relationship is to be understood, it is clearly desirable to assess the role of each partner in maintaining proximity. However, the amount one animal approaches and/or leaves the other is a consequence of its general activity as well as its tendency to remain near its partner. Plotting the frequency with which mother and infant leave each other (e.g., Kaufman & Rosenblum, 1969), or even the proportion of each partner's movements which increase or decrease proximity (Jensen, Bobbitt, & Gordon, 1967), gives no indication of the relative role of each animal in maintaining proximity. There is no wholly satisfactory solution to this problem, but the measure used here (the difference between the percentage of approaches and the percentage of leavings due to the infant; see Fig. 1) assesses the relative roles of mother and infant, and is less dependent on absolute activity levels than any of the other possibilities (Hinde & Atkinson, 1970).

4. Assessing Changes or Differences in Mother–Infant Relationships

To understand changes in, or differences between mother–infant relationships, it is necessary to tease apart the roles of the two partners in producing them. The finding that one partner is primarily responsible for, say, maintaining proximity at a given age is no indication that it is responsible for any age-changes in proximity that may be occurring around that age. A method, applicable not only to age changes, but also to differences between individuals (whether or not produced by experimental treatment), depends on a consideration of the relationships between measures.

A convenient starting point is to consider how far observed changes in the mother–infant relationship can be understood as effects of one or more of four simple types of change in the participants, namely, an increase or decrease in the tendencies of mother or infant to respond positively (other than by avoidance or aggression) to the other (Hinde, 1969). The hypothesis that changes in the relationship depend on only four types of basic change will of course prove too simple, but it is fruitful to see how far it will interpret the data.

Table III shows predicted directions of the effects of these simple basic changes on some of the measures of mother–infant relations.[2] The pattern of prediction shows that, if the changes in mother–infant interaction could in fact be accounted for in terms of one or more of these simple changes, then certain measures would always be correlated with each other. *Time off* would be positively correlated with time at a distance from the mother (>2 *feet only*); the relative frequency of rejections ($R/A + M + R$) would be positively correlated with the infant's role in maintaining proximity ($\%Ap - \%L$); and these last two would be negatively correlated with the mother's initiative in nipple contacts ($M/A + M$). The extent to which the correlations between these measures approach unity thus provides a measure of the extent to which the postulated basic changes, each affecting all measures in the manner indicated, will in fact account for the changes in the relationship.

In any case, other correlations provide an indication of the extent to which changes in the behavior of mother or infant are responsible for changes in the particular measures. Thus Table III shows that strong negative correlations

---

[2] The predictions concerning $M/A + M$ require further comment. After the first few weeks, most occasions on which the mother picks the infant up are associated with external disturbance. It is therefore assumed that changes in the infant's responsiveness to the mother will affect primarily the frequency with which it goes to the mother and is accepted by her, rather than the frequency with which she initiates nipple contacts. On the other hand, changes in the mother's responsiveness to the infant are likely to affect both the extent to which outside disturbances cause her to initiate nipple contacts and her responsiveness to the infant's demands, i.e., both M and A. The predictions in the table assume that the former predominates, but this may not be universally valid.

TABLE III

Predicted Directions of Each of Four Types of Change in Mothers' or in Infants' Behavior on Measures of Mother–Infant Interaction

		Time off	$M/A + M$	$R/A + M + R$	*>2 ft only*	*%Ap − %L*
Infant ←———	Mother	+	+	−	+	−
Infant ——→	Mother	−	−	+	−	+
Infant	Mother ——→	+	−	+	+	+
Infant	Mother ←———	−	+	−	−	−

The arrows indicate a change in the behavior of the individual with which they are associated such that it seeks proximity with its partner more or less. *Time off* = number of half minutes in which infant recorded off mother as a percentage of half minutes watched. $M/A + M$ = ratio of number of times infant was picked up by mother, on mother's initiative, to total number of times it was accepted or picked up by her. $R/A + M + R$ = ratio of number of times infant was rejected to total number of times infant attempted to gain nipple or was picked up by its mother. *>2 ft only* = number of half minutes which infant spent wholly more than 2 feet from mother as a percentage of the number in which it was off at all. *%Ap − %L* = difference between percentage of approaches (from more than 2 feet to less than 2 feet) which were due to movement by infant, and percentage of leavings (from less than 2 feet to more than 2 feet) which were due to infant. + Indicates an increase , − a decrease in the measure.

between *Time off* and the relative frequency of rejections $(R/A + M + R)$, or positive correlations between *Time off* and maternal initiative in nipple contacts $(M/A + M)$, indicate that the infant is primarily responsible for the change in *Time off*, while positive correlations in the first case, and negative in the second, that the mother is. Similarly, a negative correlation between time at a distance (> *2 feet only*) and the infant's initiative in maintaining proximity (*%Ap − %L*), suggests that the infant is primarily responsible for changes in the former, and vice versa. Arguments of this sort will be applied in a number of contexts in later sections.

5. THE NATURE OF AGE CHANGES IN THE MOTHER–INFANT RELATIONSHIP

The increasing independence of the infant monkey is correlated with, and seems at first sight to be due to, its increasing physical stature and capacities, and its increasing interest in its environment. However, three lines of evidence indicate the important part played by the mother in promoting infant independence. The first comes from a comparison of infants raised in the laboratory by their mothers with infants raised on cloth-covered surrogates. The frequency of body contacts between mother and infant decreases with age more slowly in the latter than in the former. This is presumably due to the absence of negative responses by the inanimate mother surrogates (Hansen, 1966). The extreme persistence of the clinging response in infant rhesus raised in groups of two or four, without mothers or surrogates, also suggests that maternal rejection is normally important in the waning of clinging. In peer groups the infants cling to each other and are seldom rejected (Harlow, 1969).

Evidence for the importance of the mother's role comes also from the frequency with which she rejects, hits or punishes her infant during the period in which it is becoming more independent. Thus Hansen (1966) emphasized the frequency with which caged rhesus mothers punished their infants, and the active role played by the mother in the emancipation of the infant (see also Fig. 1). Jensen, Bobbitt, and Gordon (1969), studying caged pigtailed monkeys, analyzed the sequences of behavior involving hitting. Hitting was more common in a "privation" environment than in a relatively "rich" cage environment (see Section IV, A, 7). In both environments, it led to a reduction in "climbing on" behavior. In the rich environment, hitting was usually followed by the infant leaving the mother. In the privation environment, the infant often remained on the mother but took a quieter or more dependent posture.

However, such data relate primarily to the mother's role in determining the nature of the relationship at any given time, and not to its changes with age. Here an analysis of the data in Fig. 1 by the method described in the previous section is helpful. Rank-order correlation coefficients between the medians of measures for all infants observed are shown in Table IV. Since visual inspection indicated that the relationships changed with age, separate coefficients were calculated for Weeks 1 to 6, 7 to 20 and 21 onwards. Of the pairs of measures for which the simple model of Table I predicts high correlations, *Time off* was positively correlated with time at a distance ($>2\ feet\ only$) at all ages. The relative frequency of rejections ($R/A + M + R$) was positively correlated with the infant's role in maintaining proximity ($\%Ap - \%L$), but the coefficient was significant only for older infants, suggesting that rejection by the mother and the infant's initiative in maintaining proximity are partially independent, especially in the younger age-range. The infant's role in maintaining proximity was negatively correlated with maternal initiative in nipple contacts ($M/A + M$) in the early weeks, but not later: after Week 21, $M/A + M$ was, in any case, nearly zero (Fig. 1). [The absolute and relative frequencies of rejections (R and $R/A + M + R$), while strongly positively correlated in Weeks 7 to 20, were not later. This was because the absolute frequency of rejections fell after the end of the first year, presumably because of lowered demand by the infants.]

Turning to measures that throw light on the relative roles of mothers and

TABLE IV

Rank Order Correlation Coefficients between Median Values of Measures
for All Individuals in Each Age Span

	M/A + M	R/A + M + R	>2 ft only	%Ap − %L
Time off	−0.95†	− −	+0.98†	+0.96†
	−0.85*	+0.54	+0.96†	+0.94†
	−0.53*	+0.69†	+0.74†	+0.38*
M/A + M		− −	−0.88*	−0.73
		−0.40	−0.74*	−0.86*
		−0.42	−0.07	+0.07
R/A + M + R			− −	− −
			+0.50	+0.36
			+0.57†	+0.63†
>2 ft only				+0.79
				+0.93†
				+0.57†

From upper left to lower right figures refer to weeks 1-6 ($N = 6$), 7-20 ($N = 7$) and weeks 21 on. * and † indicate $P < 0.05$ and 0.01 respectively. Rejections were too few to be assessed in the earliest period. See Table III for definitions.

infants, the negative correlations between *Time off* and the maternal initiative in nipple contacts (*M/A + M*), and the positive correlations between *Time off* and the relative frequency of rejections (*R/A + M + R*), show that changes in the mothers were primarily responsible for the increase in *Time off*; and those between *> 2 feet only* and the infant's initiative in maintaining proximity (*%Ap − %L*) show that it was again changes in the mothers that were primarily responsible for the increase in the time that the infants spent at a distance from them. Thus, although the primary roles of the mother are of course protective and nutritive, she also promotes the increasing independence of the infant. Even in the first 20 weeks, when it was the mothers who were primarily responsible for maintaining mother–infant proximity (*%Ap − %L* negative), it was not changes in the infants' behavior that were primarily responsible for their growing independence but changes in the mothers'. This of course does not imply that the changes in the mother's behavior arise endogenously. They may have been initiated by changes in the infant's behavior, such as their demand for milk or locomotor activity. These are consequences of development, which, in turn, depend on the mothers. The change in the relationship with age involves complex interactions between infant and mother.

Similar conclusions were reached by Jensen, Bobbitt, and Gordon (in press) in their study of pigtailed macaques. For example, as the infants became older, the mothers directed less of their behavior toward the infants, and more to the environment; and an increase in the proportion of the mothers' locomotion that involved leaving the infant paralleled the decrease in proximity between them.

The precise role of maternal rejections and aggression has, however, been the subject of some dispute. Kaufman and Rosenblum (1969) note that the time that bonnet and pigtail infants spend on their mothers *increases* slightly and temporarily around the age at which nipple withdrawal and punitive deterrence are most common. Comparing this with the earlier finding that air blasts from a surrogate mother caused infant rhesus to cling more tightly (Rosenblum & Harlow, 1963), they argue that maternal rejection serves to increase rather than to decrease the infant's dependent behavior. In harmony with this, van Lawick-Goodall (1968) describes increased spatial proximity between a mother chimpanzee and its offspring after weaning. Whereas maternal rejection may increase the infant's demands in the short term, especially if the infant is not yet ready for weaning, the effect is normally temporary. Furthermore, as the data cited above on hitting by pigtailed mothers show, the consequences of hitting vary with the environment: in a "rich" environment it is likely to cause infants to leave, but in a deprived one to cling quietly (Jensen, Bobbitt, & Gordon, 1969).

Kaufman and Rosenblum also argue that, since bonnet infants are more independent of their mothers that are pigtail infants, although they are rejected by their mothers less, maternal rejection behavior cannot be an important determinant of the growing independence of the young. By their own showing, however, complex social factors affecting both mother and infant influence the differences in the mother–infant relationship between the two species, and such a cross-species comparison is hardly a sound basis for deductions concerning the operation of factors within a species.

In summary, the growth of independence is of course a complicated matter, and not the result of any single cause. The increasing tendency of the infant to leave its mother to explore its physical environment and to play with its peers is undoubtedly important, but the mother also plays an important part in both permitting and promoting this. Furthermore, the consequences of the mother's actions are not necessarily so simple as would appear: hitting may lead to a temporary increase in attachment instead of detachment, and which response predominates depends in part on the environment, and in part on the infant's maturity.

6. DIFFERENCES AMONG MOTHER–INFANT PAIRS

If the nature of the mother–infant relationship has any significant influence on the production of individual differences, there must be some consistency in the differences among individual mother–infant relationships. To assess this, the coefficients of concordance for particular measures of individual mother–infant pairs over successive age-periods were calculated for rhesus living in captive groups. Considerable consistency was shown (Hinde & Spencer-Booth, 1967a).

That being the case, we may ask whether differences among mother–infant pairs are primarily due to mothers, to infants, or to both. A similar method to that just used for assessing age changes could be used here, rank-order correlation coefficients being calculated in this case among individual mean scores over particular age periods (Table V). In brief, the analysis indicated that before Week 20 the factors controlling the time the infant spent off its mother and at a distance from its mother were related, but with older infants there was considerable independence among the factors. The correlations between *Time off* and the relative frequency of rejections ($R/A + M + R$), and between time at a distance (>2 *feet only*) and the infant's role in maintaining proximity

TABLE V

Correlation Coefficients between Individual Mean Scores for the Periods
(upper left to bottom right) 1-6 Weeks, 7-12 Weeks, 13-18 Weeks, and 20 Weeks and Later

	$M/A + M$	$R/A + M + R$	>2 ft only	$\%Ap - \%L$
Time off	−0.56* −0.30 −0.30 −0.28	− − +0.41 +0.13 −0.44	+0.62† +0.59† +0.79† +0.35	+0.04 +0.50* −0.14 +0.01
$M/A + M$		− − −0.42 −0.36 −0.07	−0.74† −0.53* −0.58* −0.05	−0.90† −0.53† −0.06 +0.05
$R/A + M + R$			− − +0.72† +0.26 +0.20	− − +0.84† +0.37 +0.49
>2 *ft only*				+0.26 +0.69† −0.11 −0.01

In the first three cases the coefficient is based on 14-22 individuals. In the last case the figure is the median coefficient for 8 individuals for all 6 or 8 week periods up to week 132. See Table III for definitions.

$(\%Ap - \%L)$, although never high, change from positive to negative with increasing age, while those between *Time off* and the maternal initiative in nipple contacts $(M/A + M)$ become decreasingly negative. This indicates that individual differences in mother–infant interaction are due to differences in both mothers and infants, but that the differences among mothers are more important with young infants, and those among infants with older ones (Hinde, 1969; Hinde and Spencer-Booth, 1971a).

7. INFLUENCE OF PHYSICAL ENVIRONMENT ON THE MOTHER–INFANT RELATIONSHIP

In many of the Wisconsin studies of rhesus development, the infants could leave their mothers through doors too small for the mother to follow. Clearly, such gross factors in the physical environment may markedly affect the mother–infant relationship.

More subtle effects of the physical environment were studied by Jensen, Bobbitt, and Gordon (1968b) in pigtailed macaques over the first 15 weeks of life. Two environments were used: a "privation" environment consisting of a bare wire cage in a soundproof room, and a so-called "rich" environment, which provided opportunity for climbing and manipulation, and visual and auditory stimulation from other monkeys. Infants in the rich environment spent more time at a distance from their mothers (Fig. 5), and oriented more behavior towards the environment and less towards themselves. The privation environment resulted also in motor retardation of the infants. This study suggests that complexity of the physical environment may have a considerable effect on the mother–infant relationship.

Spencer-Booth (personal communication) also found differences between infants living in small cages and infants living in groups in larger cages. In particular, a higher proportion of the former's attempts to gain the nipple were rejected. Rowell (personal communication) studied Sykes' monkeys living in a group in a cage from which the infants could escape. At 6 months of age the infants were relatively rarely on their mothers, but when the exit was sealed they reverted to clinging and sucking like babies.

The bases of these effects of the physical environment have not yet been studied. However, Mason (1967), in a series of experiments on chimpanzees, claims that the mother–infant relationship depends in part on the reward values for the infant of contact, play, and being groomed. He suggests that these rewarding effects are mediated in part by changes in "arousal." If environmental factors compete with or complement those provided by a social companion, the relationship with the companion will certainly be affected.

8. ROLE OF MATERNAL EXPERIENCE

It is often suggested that primiparous mothers are less efficient than more

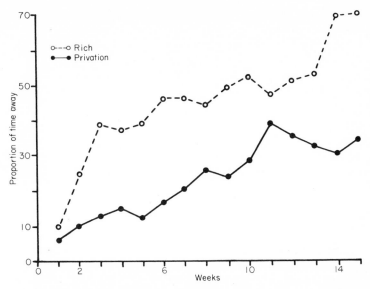

FIG. 5. Proportion of time mother and infant pigtailed monkeys spent separated in "rich" and "privation" environments. (After Jensen, Bobbitt, & Gordon, 1968b)

experienced ones, but there is little hard data. Seay (1966) suggests that some of the earlier records of inadequate mothering by primiparous chimpanzees could have been related to the conditions of captivity. His own records of captive rhesus monkeys, and van Lawick-Goodall's (1968) data on free-living chimpanzees, show that primiparous mothers normally give adequate maternal care. It remains likely that practice brings improvement (e.g., Jay, 1963). Surrogate-reared rhesus, who showed markedly inadequate maternal behavior when their first infants were born, were much more competent with their second (Harlow & Harlow, 1969).

9. SEX DIFFERENCES

Both in rhesus (personal observations) and in bonnet and pigtailed macaques (Rosenblum & Kaufman, 1967), the mothers display particular interest in the scrotum of male newborns, peering at it and fingering it. In baboons, similar attention is directed to the penis (Rowell, personal communication). Mitchell (1968a) has demonstrated behavioral differences between male and female rhesus infants, and between maternal responses to infants of the two sexes, even in the first week. In pigtailed monkeys mother–male-infant pairs are at first more mutually dependent, but within a few weeks show a rapidly accelerating trend towards greater independence, than mother–female-infant pairs. This fact seems to be immediately due to differences between maternal responses to male and female infants (Jensen, Bobbitt, & Gordon, 1968a; see also Hinde & Spencer-Booth, 1971a).

While male rhesus infants are more aggressive and self-assertive than females (see Section IV, C) they are more damaged by rearing in total or partial social isolation than are females (Sackett, 1968b). They are also somewhat affected by a period of maternal deprivation when 21 to 32 weeks old (Spencer-Booth & Hinde, 1971).

In other species, there are indications of greater independence among males than among females. Thus male hamadryas baboon infants leave their mothers more than do females (Kummer, 1968b). In many other species, the males spend less time near their mothers after weaning than do the females (see also Section IV, D).

10. Some Long-Term Consequences of the Particular Nature of the Mother–Infant Relationship

It is clear that absence of mother, or peers, or both can markedly affect an infant's behavioral development. If social companions are present, we may ask to what extent the particular character of the relationship with them affects the infant. As yet, only two groups of studies, both concerned with highly abnormal situations, are available.

The Wisconsin rhesus monkeys reared under conditions of social deprivation became very inadequate mothers, behaving brutally or with indifference to their offspring. Their infants tended to be inferior to normally reared controls in the development of play and sex behavior (Fig. 6), and significantly more aggressive. Some of the animals were tested when 3 years old and still showed some

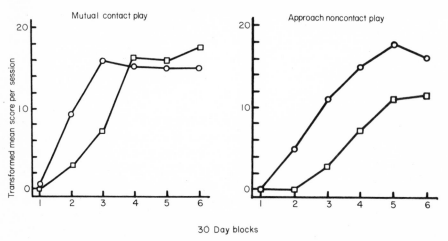

FIG. 6. Development of contact and noncontact play by infant rhesus with motherless mothers (squares) and feral mothers (circles). (After Harlow & Harlow, 1969)

deficiencies (Seay, Alexander, & Harlow, 1964; Arling, 1966, cited by Harlow & Harlow, 1969; Mitchell, Raymond, Ruppenthal, & Harlow, 1966; Mitchell, 1968b; Harlow, 1969). A major source of difference between these motherless–mother infants and animals with feral mothers appears to lie in the utilization of signal movements (Møller, Harlow, & Mitchell, 1968).

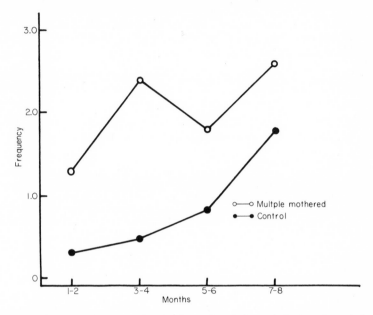

FIG. 7. Disturbance behavior of multiple-mothered rhesus infants and controls. (After Harlow & Harlow, 1969)

The second study concerns a comparison of infants raised by their own mothers with four infants who were rotated among mothers on a biweekly schedule. Although the mothers in the rotated group showed apparently normal maternal behavior, the infants showed more "disturbance" behavior (Fig. 7): they did not, however, show abnormalities in play, social, or sexual behavior (Griffin, cited by Harlow & Harlow, 1969). At 3 years of age, these animals tended to be higher in social dominance than normally reared animals (Sackett, 1968a).

As yet, unfortunately, there are no studies of the extent to which individual differences in temperament among "normal" mothers affect the development of their infants. Field studies suggest that such effects would be worth pursuing. Thus Jay (1962, 1963) found that experienced multiparous langurs handled their infants in a more competent manner than inexperienced mothers, and were less likely to startle their infants by sudden movements. Mitchell, Raymond,

Ruppenthal, and Harlow (1966) have demonstrated behavioral differences between the 30-month-old offspring of multiparous and primiparous rhesus mothers. The former showed more hostility, play, and self-manipulation, and less fear of adults, stereotyped movements, and disturbances. Again, DeVore (1963) found that baboon infants with dominant mothers were subject less to insecurity and frustration, and that subordinate mothers were shorter-tempered and less responsive to their infants. Fossey (personal communication) reports a similar difference in the gorilla. However, precise data on such dominance-related differences in maternal behavior, or on their consequences on the offspring, are not yet available. In the absence of further experimental data, we may consider some contexts in which the infant's subsequent behavior is determined by that of the mother, or vice versa.

On the whole, the mothers of nonhuman primates seem not to teach their infants. In a number of species, a mother has been seen to move a little away from her infant and then to wait while it crawled after her (e.g., Howler monkeys, Carpenter, 1934; rhesus, Hinde, Rowell, & Spencer-Booth, 1964; gorilla, Schaller, 1963; chimpanzees, van Lawick-Goodall, 1968): this has the effect of encouraging the infant to walk, but can hardly be called teaching. However, it is clear that infants learn a great deal from their mothers, especially in the context of avoidance and food-getting behavior. Even avoidance of snakes differs between laboratory and wild-reared monkeys, and may depend in part on parental example (Joslin, Fletcher, & Emlen, 1964). It has been shown in the laboratory that monkeys can learn to avoid situations or responses that are seen to cause pain to other individuals (Child, 1938; Hansen & Mason, 1962; Hall, 1968), and to accept food that other individuals are seen to take (Weiskrantz & Cowey, 1963). In nature, the infant's proximity to its mother ensures that it becomes rapidly conditioned by her fear responses (e.g., Baldwin, 1969), and that its feeding behavior is influenced by her (e.g., Hall, 1962). In the patas monkey (Hall, 1965), Japanese macaque (*Macaca fuscata*) (Kawamura, 1959), and chimpanzee (van Lawick-Goodall, 1968), the young eat fragments that their mothers drop, as well as being especially likely to feed at the same food sources. Although by the time they are 1 year old, Japanese macaques are acquainted with all the types of food used by the troop, it is difficult to make them take new types of food in the laboratory. Apparently learning from the mother is normally important (Kawamura, 1959). Schaller (1963) records an infant gorilla removing food from its mother's mouth and eating it, and one case of a mother breaking off a stem for its infant to eat. Imitation, principally of the mother, is important for the development of tool-using behavior in wild-living chimpanzees (Goodall, 1964; van Lawick-Goodall, 1968); and the development of actions by imitation has also been recorded in hand-reared individuals (Hayes & Hayes, 1952; Kellogg, 1968). In the latter case, the actions may be used for social communication (Gardner & Gardner, 1969, 1971).

In squirrel monkeys, food-catching skill is learned by younger juveniles from older ones, rather than from their mothers (Baldwin, 1969). However, it is by no means always the younger animals that learn food habits from older ones. Under natural conditions, young animals investigate new objects more than do older individuals, and this may lead to a transfer of feeding habits from younger to older animals. Thus, among the Japanese macaques, new foods tended to be accepted first by juveniles, and their use then diffused through the colony via their mothers and then the mothers' younger offspring and consorts (Itani, 1958). Although diffusion sometimes occurs in the opposite direction (Frisch, 1968), kinship ties are probably always important (Kawamura, 1959; Tsumori, 1967). The learning processes involved are discussed by Hall (1968). Comparable processes have been studied in the laboratory by Mason (1959), Hansen and Mason (1962), and Mason and Hollis (1962).

In a number of species the mother's dominance status affects that of her offspring. Thus in the Japanese macaque the offspring of the higher-ranking mother is usually successful in competition over food items (Kawai, 1958, cited by Jay, 1965). Imanishi (1960) has suggested that a young male's entry into the center of a group requires his acceptance by the dominant females, and that this is influenced by the rank of his mother (see also Yamada, 1963). In the rhesus monkey, Koford (1963a, b,) found that the sons of high-ranking mothers tended to be high in the adult male hierarchy, and Sade (1967) found that young rhesus monkeys start to fight as old infants or young yearlings, defeating those age peers whose mothers rank below their own. The hierarchies thus established persist for several years, and as the females become adult, they come to rank just below their mothers in the hierarchy of adults. Sade suggests that maternal rank may be less important in determining the ultimate status of males. Marsden (1968), studying a captive group, found that experimental alteration of the mothers' ranks was nearly always associated with a similar change in the infants'. Marsden emphasizes that the offspring were not always merely passive partners in a dominance change, and could contribute to a rise.

In these macaques, therefore, an influence of the mother's rank on that of her infant is well established. Since the stability of the social structure is related to the maintenance of the dominance hierarchy, the absence of kinship ties may cause instability in artificial groups (Vandenbergh, 1967). Similar effects no doubt occur in other species (e.g., baboons, DeVore, 1963), especially in ground living ones, though other influences no doubt also operate (e.g., patas monkey, Hall, 1967).

11. TWINS

Twins are extremely rare in most monkeys and apes, and only one study of behavioral development, of twin rhesus infants born in a cage environment, is available (Spencer-Booth, 1968b). The twins were distinguishable in appearance, and one was more active than the other from an early age. The more active one spent less time asleep and on the nipple, and more time off and at a distance from the mother, than the less active one.

Considerations of the type discussed in Section IV, A, 4 indicated that the greater time that the more active infant spent off the mother was primarily a consequence of a difference in the behavior of the mother to the twins. She groomed it less and hit it more, and protected the other infant from it. These differences in maternal behavior may in turn have been consequences of the differences in activity between the twins.

The greater time that the more active twin spent at a distance from its mother was, however, more immediately due to differences between the twins. It was not clear to what extent the smaller tendency of the more active twin to keep close to its mother ($\%Ap - \%L$) was due to punishment by the mother (Spencer-Booth, 1968b).

12. THE EFFECTS OF TEMPORARY MOTHER–INFANT SEPARATION

A brief period of maternal deprivation may have far-reaching consequences on the personality development of the human infant (Bowlby, 1951, 1969; Ainsworth, 1962). The possible occurrence of similar effects in monkeys has been investigated in several studies. Some of these were concerned with the immediate consequences of brief separations (Jensen & Tolman, 1962), others with the effects of deprivation lasting some days or weeks on mother–infant interaction during the next few weeks (Seay, Hansen, & Harlow, 1962; Seay & Harlow, 1965; Kaufman & Rosenblum, 1967, 1969), and others with studies in which both immediate and long-term consequences were examined (Spencer-Booth & Hinde, 1967, 1971; Hinde & Spencer-Booth, 1971b). The studies differed also in the living conditions (e.g., social groups, single mother–infant pairs with limited opportunity for peer interaction), and in the age at separation and its duration.

Figure 8 shows data for group-living rhesus infants whose mothers were removed from the group for 6 days when the infants were between 21 and 30 weeks old (Spencer-Booth & Hinde, 1971). Except as specified in the legend the figure shows, for each behavioral characteristic, the group median (filled circles and thick lines), and, since individual differences were considerable, the two extreme individual cases chosen on the basis of preseparation scores. Soon after the mother was first removed the infants showed extreme distress, including "whoo calls" and screaming. They then showed reduced activity, spent more time sitting, and played less. (Kaufman and Rosenblum, on the basis of their studies of separated pigtailed infants, regard the symptoms as similar to the anaclitic depression of human infants.) As the separation period progressed, the rhesus infants became active for more of the time, though less active when active. (Kaufman and Rosenblum, who continued some of their separations for a month, found that the depression abated after a while, so that by the time the mother was returned, some infants seemed almost normal.)

When the rhesus mothers were returned, the intensity of the mother–infant relationship was much increased. Nearly all infants spent less time off and at a distance from their mothers than they had before separation. By the arguments discussed in Section IV, A, 4, this was due primarily to the intensity of demand by the infant, the infant's role in maintaining proximity ($\%Ap - \%L$) being increased. During the following days there were considerable fluctuations in the nature of the mother–infant relationship, revealed especially in the measures of time off the mother and the relative frequency of rejections, as mother and infant readjusted to each other.

During the mother's absence, many infants spent considerable time on or near other group companions, though this study of rhesus infants provided little evidence that this ameliorated the effects of the maternal deprivation. [Kaufman and Rosenblum ascribe the absence of depression in one of the pigtailed infants they studied to the fact that, unlike the others, it was able to derive comfort from contact with other adults. Likewise they found the effects of maternal deprivation to be much less severe for bonnet infants, which interacted freely with other adults (see Section IV, B), than for pigtailed infants.]

As Fig. 8 shows, there were wide individual differences in the short term effects of a period of maternal deprivation. These were correlated with the nature of the mother–infant relationship. Those infants, which were rejected most by their mothers and played the greatest part in maintaining mutual proximity, showed most "distress" (assessed by frequency of whoo-calling, lack of locomotor activity, etc.,) on the mothers' return (Hinde & Spencer-Booth, 1971b).

There is some disagreement over the longer-term effects of a period of maternal deprivation. Seay and Harlow (1965) found that 2 weeks of maternal deprivation had "transient and apparently unimportant" effects on the mother–infant relationship. Kaufman and Rosenblum (1967) found that 1-month-long separations produced no obvious interference with the subsequent development of pigtailed monkeys though there were suggestive indications of subtle effects. Spencer-Booth and Hinde (1967, in preparation), however, found that some of the rhesus infants whose mothers were removed for 6 days (see Fig. 8) showed persisting aftereffects. For example, the first few infants tested differed significantly from eight controls which had had no deprivation experience in tests given at 12 and 30 months of age (see Section IV, A, 13).

In a study of a somewhat different type, Mitchell, Harlow, Griffin and Møller (1967) found that four rhesus infants which, for the first 8 months of their lives, were separated from their mothers for 2 hours every fortnight and then returned, gave more whoo vocalizations and showed more fear than non-separated controls. At 3 years of age, animals previously subjected to repeated separations were low in social dominance as compared with normal animals

(Sackett, 1968a). That the depression produced by maternal deprivation is not an artifact of laboratory conditions, and that it occurs also in apes, is indicated by van Lawick-Goodall's data on wild-living chimpanzees. Three 1- to 3-year-old infants, whose mothers died, became depressed and two of them died within a year, even though one of the latter was adopted by an elder sibling.

Less attention has been paid to the effect of separation from their infants on mothers, but Jensen (1968) found that five pigtailed mothers showed first agitation and then (at 18 days) a state resembling depression. By 2 months all measures had returned to preseparation levels.

FIG. 8. Effect of 6 days maternal deprivation on rhesus infants about 21 weeks ($N = 6$), about 25 weeks ($N = 5$) or about 30 weeks ($N = 5$) old. The abscissas are scaled in real time (the absence of the mother being indicated by a discontinuity), but data are lumped as follows:

PrS	Mean of 3 pre-separation watches
S1	Watch immediately after mother removed
S	Mean of all 4 separation watches (including S1)
PtS1	Watch immediately after mother returned
PtS2-7	Mean of all 4 watches on 2nd to 7th days after that on which mother returned
PtS14-28	Mean of all three watches on 14th to 28th days after that on which mother returned.

The data for the several age groups are lumped, each figure (except for Time spent on group companions) showing the median for all individuals (thick lines) and two extreme individuals (thin lines). The latter were the two giving the highest and lowest pre-separation means for which complete data were available, except for Whoos and Tantrums, for which the two individuals giving the extreme separation means were used. Since medians were mostly zero for Time on group companions, two extreme and one intermediate individuals are plotted. Data for only one of the 30-week individuals were available for Leaning against group companions.

Ordinates. (a) *Total time off mother.* Number of half minutes in which recorded off mother as percentage of half minutes watched. (b) $R/A + M + R$. Number of times infant's attempts to gain nipple rejected by mother, divided by total number of attempts plus number of times it gained nipple by mother's initiative. (c) >2 *feet.* Number of half minutes in which recorded more than 2 feet from mother as percentage of number in which it was off all animals. (d) $\%Ap - \%L$. Difference between percentage of approaches due to infant and percentage of leavings due to infant. (e) *Time on group companions.* Number of half minutes in which recorded as percentage of half minutes watched. (f) *Time leaning against group companions.* Number of half minutes in which recorded as percentage of number of half minutes off all animals. (g) *Whoos.* Number per 100 minutes in which infant off all animals. (h) *Tantrums.* Number of half minutes in which recorded as percentage of number in which infant recorded off all animals. (i) *Activity.* Mean number of cage sections entered per half minute off all animals (cage divided into 16 sections). (j) *Sitting.* Number of half minutes in which recorded as percentage of number in which recorded off all animals. (k) *Manipulative Play* and (l) *Rough & Tumble Play.* Number of half minutes in which recorded as percentage of number in which recorded off all animals. (After Spencer-Booth & Hinde, 1971.)

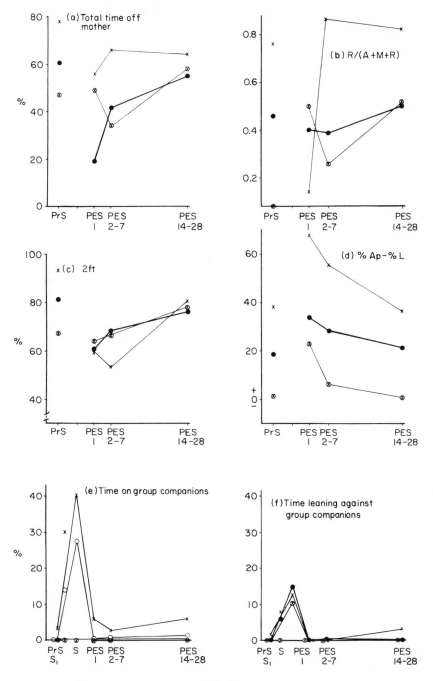

FIG. 8

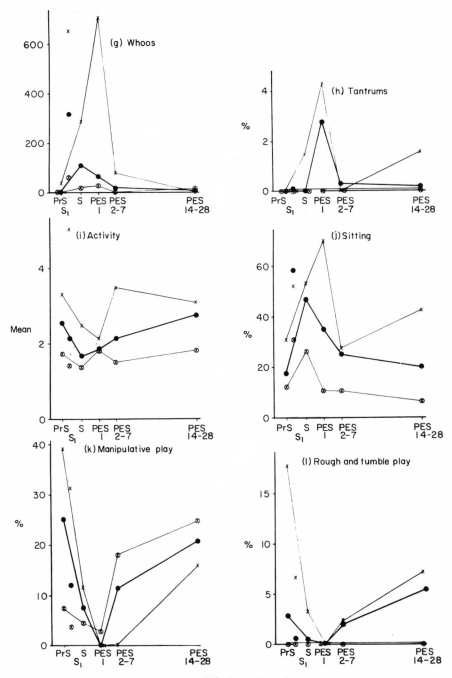

FIG. 8—continued

13. Assessment of the Effects of the Social Environment
    (Including Maternal Deprivation) on Behavior

The Wisconsin studies have shown that rearing monkeys under conditions of gross social deprivation can produce marked distortions in their adult behavior as, for instance, inability to mate or rear infants (Section III). These shortcomings were, in many cases, foreshadowed by their social behavior with peers in a playpen situation, or their performance in a visual exploration apparatus.

The next stage requires a search for more subtle effects—means of measuring such individual differences in temperament as natural variations in the social environment might produce. One possibility is to measure social interactions in a continuous group-living environment. As seen above (Section III), such a method can show that some of the effects on mother–infant interaction of living in social isolation persist in group conditions. It would be preferable, however, to use more standard testing conditions than are possible in a social situation. To this end, Spencer-Booth and Hinde (1969) are assessing the usefulness of a variety of tests, mostly involving responses to strange objects and mildly disturbing or frustrating situations. These have been used to compare infant rhesus macaques, reared by their mothers but otherwise socially isolated, with control infants reared in groups. At 6 months, the isolated infants came off their mothers sooner when presented with a strange object, and at 12 months responded less strongly to mildly disturbing situations, than the group-reared infants. The tests also revealed differences between 12-month-old infants that had been separated from their mothers for 6 days at 30 to 32 weeks and nonseparated controls. The previously separated infants were less ready to take vitamin pills from an experimenter, showed greater initiative in maintaining proximity to their mothers when moved to an indoor cage, and visited an adjoining cage containing strange objects less than controls that had had no separation experience. Some significant differences were also found in tests at 30 months. The separated infants were less active after being frightened, and less ready to take vitamins or bananas from an experimenter. While a large number of the tests used did not give significant differences (Experimentals $N = 4$ and Controls $N = 6$), there was none that gave them in a direction suggesting that the controls were more disturbed by slightly strange situations than the separated infants. Further data confirm these conclusions (Spencer-Booth & Hinde, in preparation).

B. Female Group-Companions

This section is concerned with the behavior shown toward infants by adolescent and adult females other than their own mothers,[3] and maternal-type responses shown by females of all ages. Here again, the rhesus has been studied more intensively than any other species.

1. RHESUS MACAQUES

Female group-companions show interest in rhesus infants in both wild troops (Southwick, Beg, & Siddiqi, 1965; Kaufmann, 1966, 1967) and captive groups (Rowell, Hinde, & Spencer-Booth, 1964). The behavior they direct toward the infants includes investigation by visual inspection and touch, grooming, protection, cuddling and carrying, play, and aggression. The mother attempts to protect her infant from such attentions, either by removing it or threatening off the aunts. If it were not for this, it seems certain that carrying and cuddling would be more prominent. When mothers were removed from captive groups, some, but not all, of the deprived infants spent much time in contact with aunts, and in some cases were held or carried by them for considerable periods (Spencer-Booth & Hinde, 1967, 1971). In some circumstances, infants may be adopted by childless females (Hansen, 1966; Rowell et al., 1964). Indeed, Sackett (personal communication) states that females in the Wisconsin colony may adopt a baby even while still caring for their own. He regards such behavior as no more common in females without infants than in females with infants.

The extent to which females interact with infants varies with their age and reproductive status. Data from captive groups (Spencer-Booth, 1968a) showed that females 2 years older than the infants (which had no young of their own) interacted with infants that were less than 30 weeks old more than did females only 1 year older than the infants, and also more than females more than 2 years older than the infants (Fig. 9; see also Kaufman, 1966). Females that had borne young of their own interacted with infants less than nulliparous adults, the difference being especially marked (except for aggressive responses) if they had young at the time. However, females with young of their own showed more aggressive responses to infants than did any other age–sex category. Where an infant had both a sibling and a nonsibling in the same age–sex category present, the former nearly always interacted with it more. This was especially so with

---

[3] Adolescent and adult female group-companions can be termed "aunts," with the proviso that this carries no implication of blood-relationship (e.g., Rowell et al., 1964). Raphael (1969) objects to the term because of the implication of kinship ties, and it might be better to drop the term than to start a dreary terminological argument. However, Raphael's suggestion that the Greek term "doula" be substituted is likely to raise other problems, especially as she stipulates that doula behavior "should not appear to cause a negative response in the recipient". This would make it inapplicable in many instances of interaction between infant monkeys and adult females, as well as to human aunt/infant relations. "Aunt" is therefore sometimes used here as a convenient shorthand.

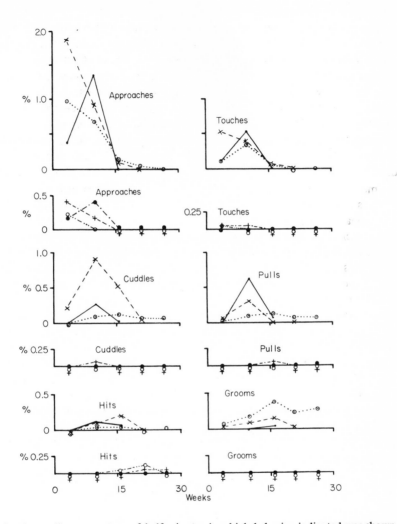

FIG. 9. The median percentage of half minutes in which behavior indicated was shown towards infants by various categories of female group companions. (After Spencer-Booth, 1968a)

●————●	Females about 1 year older
x – – – – – x	Females about 2 years older
⊙·········⊙	Females >150 weeks older which not had live young
● –·–·–·– ●	Females >150 weeks older with live young either not present or at least 2 years old
○ – ··· – ··· – ○	Females >150 weeks older with young of same crop present
+ – – – – – +	Females >150 weeks older with young of previous crop present

young infants, and was probably a consequence of the greater permissiveness of the mother towards the sibling (Spencer-Booth, 1968a).

The nature and extent of aunt–infant interaction also varies with the age of the infant. This variance is due principally to three factors that vary with the infant's age. First, its stimulus characters vary. Their effectiveness in eliciting maternal-type behavior is probably greatest in the early weeks, and declines with age. Many other species show a change in coat color, which may decrease the infant's attractiveness. In rhesus monkeys, the loss of the central parting in the head hair may be important. Their effectiveness in eliciting play or adult social responses increases later. A second issue is the behavior of the infants, for they are not merely passive partners in such interactions. If their mothers are present they may actively avoid the aunt's attentions, and fail to cling if picked up: if the mother is absent, however, they often actively solicit attention from aunts. (Infants are clearly able to recognize their own mother when a week or two old, and perhaps even earlier.) A third factor is the permissiveness of the mother, which increases with the age of the infant. In the very early weeks, aunts are permitted merely to approach the infants and inspect them visually. Touching increases somewhat later, and carrying and cuddling does not reach its peak until the second 6 weeks. Carrying and cuddling decrease after this, probably in part because the infants become less attractive and in part because they are better able to elude the aunts. Aggressive behavior and grooming are most frequent in the third 6-week period; again the decrease is probably due in part to increased powers of evasion by the infants and their increased respect for adult monkeys.

The mother's permissiveness may depend in turn on her social relationship to the aunt. In captive groups, mothers are more likely to let "best friends" approach the infants (Rowell et al., 1964). Reciprocally, aunts differentiate between infants on the basis of their relationships to their mothers, and have been seen to break up fights between babies in support of a friend's infant. The mother is also more likely to let her infant go to a subordinate aunt than to a dominant one, though an aunt may take the baby by virtue of her dominance. Of all categories of group companion, the mother is most permissive to her own older offspring, and other kinship ties may also be important.

2. OTHER SPECIES

In many other species, females show some degree of interest in infants, and may attempt to show maternal behavior towards them (e.g., squirrel monkey, Hopf, 1967; Baldwin, 1969; howler monkey, Altmann, 1959; Carpenter, 1934, 1965; olive baboon, *Papio anubis,* Rowell, personal communication; hamadryas baboon, Kummer, 1968b; vervet, Struhsaker, 1967; Booth, 1962; Gartlan & Brain, 1968; Campbell's mona monkey, *Cercopithecus campbelli,* Bourlière, Bertrand, & Hunkeler, 1969; *Cercopithecus nictitans martini,* Struhsaker, 1969;

mangabey, Chalmers, 1967; patas monkey, Hall & Mayer, 1966–1967: Barbary ape, *Macaca sylvana,* Lahiri & Southwick, 1966; Deag & Crook, personal communication; Japanese macaque, Sugiyama, 1967). Comparable data are available for apes (gorilla, Schaller, 1963; Fossey, personal communication; chimpanzee, van Lawick-Goodall, 1968). The general pattern in these species is similar to that in the rhesus, with species differences stemming largely from differences in the permissiveness of mothers. Thus Lahiri and Southwick (1966) contrast the permissiveness of Barbary ape mothers with the restrictiveness of rhesus monkeys. In their study of captive groups, Rosenblum and Kaufman (1967) found that pigtailed macaque mothers tended at first to leave the group, thwarting the interest of others in the infants by withdrawal or aggressive behavior. Bonnet macaque mothers, by contrast, returned to the group soon after the birth and frequently permitted other females to explore, handle and groom their infants. (It should be mentioned that no such permissiveness was noted in a field study by Simonds, 1965.)

In Campbell's mona monkey, Bourlière *et al.* (1969) recorded mothers leaving infants with other females when they were only 16 days old. However, an extreme case is the hanuman langur, where the mother allows female group-companions to handle the infant within a few hours of birth. Observations by Jay (1963) show that an infant may be carried by as many as eight or ten females and taken as far as 75 feet from its mother in the first 2 days of life. The mother is mostly apparently unconcerned, but at the slightest sign of danger she runs to the infant: she can take her infant from any female in the troop. No influence of dominance or intimacy between mother and aunt is evident. If, however, a baby is taken by a female from another troop, its recovery may involve a clash that includes the troop male (Sugiyama, 1965a, 1965b; Yoshiba, 1968). A similar situation obtains in John's langur (*Presbytis johnii*) (Tanaka, 1965).

In langurs, as in other species (e.g., mangabey, Chalmers, 1967, 1968), the infant often resists being taken by the aunt. The langur mother never hands her infant to the aunt, but may lean backwards to facilitate its being taken (Jay, 1963; Sugiyama, 1965b). Mothers of older langur infants may leave them playing with another female, who cares for all the infants in a play group. The aunt's interest declines around the fourth month, although adult females may play with older infants (see also Sugiyama, 1965a, 1965b, 1967; Yoshiba, 1968).

Few data on the relative roles of different categories of aunt, or on the changes in aunt behavior with age of infant, are available from species other than the rhesus, but the situation seems to be similar to that in the rhesus. In langurs, aunts that nurse the infants are most often subadult females or adult females without infants. Two- to 3-year-old animals are attracted by the infants, but are more hesitant to take them from their mothers. Female vervets 6 months to 4 years in age are more prone to interact with infants than are fully adult females

(Struhsaker, 1967). In Campbell's mona monkey, infants are often carried by adult females, but sometimes by females as little as 1 year old (Bourlière, Bertrand, & Hunkeler, 1969 and personal communication). In squirrel monkeys, Baldwin (1969) recorded that the few older adult females that were without infants often formed a stable association with a particular mother, and behaved maternally toward the infant. Often, late juvenile females also interacted with infants. At first, they tended to play roughly with infants, but apparently learned to show more adultlike maternal behavior in a few weeks.

The importance of kinship ties is seen especially in van Lawick-Goodall's data on free-living chimpanzees, where much of the "maternal" behavior received by infants from other than their own mothers came from siblings. On three occasions when a mother died, the infant was adopted by a sibling; in one case, the sibling was a male. An orphaned infant that had no sibling was not adopted.

An aunt's response to the infant may also depend on previous social acquaintance with it. Kaufman and Rosenblum (1969) found that bonnet infants whose mothers had been removed were cared for by aunts, but an infant introduced into a strange group was not.

Baboon infants seem to become more attractive over the first few days of life (Hall, 1965), though even newborns arouse intense interest, and females may touch and even carry them. Rowell, Din, and Omar (1968), studying caged animals, found a sharp decline in the interest shown by other females when the infants were only a few weeks old, though some field studies suggest that this decline occurs rather later (DeVore, 1963; chacma baboon, *Papio ursinus* Bolwig, 1959). The juveniles and older infants, however, continue to be interested in the infants and attempt to play with them (DeVore, 1963; Rowell *et al.,* 1968). In squirrel monkeys, other females start to show interest in infants only when, in the third week, the latter start to be responsive to their approaches (Baldwin, 1969).

In a number of species, the waning of interest in infants is associated with a change in external appearance, the natal coat color being different from that of the adults. Thus in langurs the interest of aunts wanes when the coat color changes from brown to white (Jay, 1962; Sugiyama, 1965a), and in baboons from black to brown (Hall & DeVore, 1965; Kummer, 1968b). The natal coat is also important in eliciting maternal responses from adults in guenons (*Cercopithecus* spp.) (Booth, 1962).

Aunts affect the infant not only by their direct interactions with it, but also by affecting the mother–infant relationship. Hinde, Rowell, and Spencer-Booth (1964) found that two mothers in a group of captive rhesus were very restrictive with their infants, apparently as a consequence of the attentions of an adolescent female who tried, often with success, to groom, hold, cuddle, and play with the infants. Hinde and Spencer-Booth (1967b) therefore compared mother–infant interaction in group-living rhesus with that in mother–infant pairs segregated in similar cages. It was found that segregated infants were off their mothers more and for longer periods, and went more than 2 feet away from them more often. Application of the arguments discussed in Section IV, A, 4 showed that the segregated mothers were more permissive than the group-living ones. The greater time the infants spent off their mothers was related to a higher relative frequency of rejections, and the greater time at a distance was related to a greater role of the infant in maintaining proximity ($\%Ap - \%L$) (Fig. 10). Indeed, in the segregated infants, $\%Ap - \%L$ was positive from an early age, indicating that even then the infant was primarily responsible for the

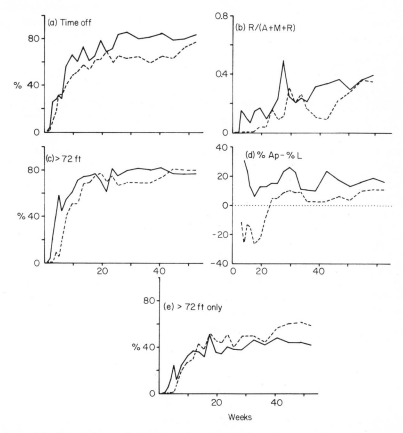

FIG. 10. Comparison of mother–infant interactions in segregated and group-living mother–infant pairs. Only medians are shown. Measures as in Fig. 1. ("$>$ 2 feet" indicates proportion of half minutes in which infant was off mother in which it went more then 2 feet from her, while "$>$ 2 feet only" indicates proportion of half minutes in which infant was off mother which it spent continually more then 2 feet from her. (After Hinde & Spencer-Booth, 1968)

maintenance of proximity. Although the segregated infants went more than 2 feet from their mothers more often than the group-living ones, after 10 weeks of age they spent fewer whole 30-second periods away. This was perhaps related to the absence of social companions: for the segregated infants, the mothers were the only social companions available (Hinde & Spencer-Booth, 1967b). Some of the differences persisted when the segregated mother–infant pairs were placed in groups when the infants were 1 year old. In addition, the previously segregated infants interacted with their group companions less than did controls (Spencer-Booth, 1969).

In this study, the animals lived in segregated or group conditions throughout the first year of life. Rowell (1968) studied the effects of short periods of segregation from the group by comparing measures of mother–infant interactions in captive group-living baboons with those obtained from the same mother–infant pairs segregated from the group or confined with one other group-member at a time. Segregation produced changes similar to those found in the rhesus study.

It may be noted here that, in a study of separately caged pigtailed mother–infant relations, Jensen, Bobbitt, and Gordon (1967) found that the infant did more approaching than leaving the mother, and mother did more leaving than approaching. This suggests that from the start, the infant was primarily responsible for maintaining proximity, and resembles the data from rhesus mother–infant pairs kept separately, but not those from group-living ones.

C. Peers

In all species of monkeys and apes, the infant is normally born into a group consisting of both adults and other infants. Since most species have a fairly well-marked birth season, there are usually a number of infants closely similar in age. Interaction with these other infants takes the form mainly of "play," a category of behavior that is impossible to define, but inescapable in practice (e.g., Hinde, 1970). Play behavior in nonhuman primates has been reviewed recently by Mason (1965b) and Loizos (1967).

1. RHESUS MACAQUES

Harlow and Harlow (1965, 1969) and Harlow (1969) describe the development of the peer–peer "affectional system" as comprising four stages, comparable to four stages in inanimate object play. Their description depends on the behavior of infants—usually without mothers—in a small "playroom." In the first "reflex" stage, infants of up to 20 to 30 days old move about in close physical proximity, and the authors suggest that this develops from exploratory responses elicited by nearby objects. In our group situation, infants of this age often orient and advance toward other infants, but the close association that the Harlows describe is almost invariably prevented by the mothers. Our data, therefore, would give no solid basis for distinguishing this from the Harlows' next "exploration" stage, in which exploration of physical objects merges into exploration of social companions. Indeed, the infant's very first exploratory responses are made to parts of its mother's body. The next stages of "interactive" and "aggressive" play involve "approach–withdrawal," which involves chasing but minimal physical contact, and "rough-and-tumble" play, which merges into aggressive play in which the bites may be painful. The Wisconsin data (plotted as "mean score per subject") indicate that rough-and-tumble and approach–withdrawal play develop in parallel and are more or less equally frequent. Data from captive groups containing infants of varied age indicated that rough-and-tumble play was more frequent than approach–withdrawal, especially during the first 6 months, but this may merely reflect the inability of younger animals to avoid the approaches of older ones (Hinde & Spencer-Booth, 1967a).

Threatening behavior and rough-and-tumble play are more common in laboratory-reared male than in female rhesus infants. Females are more likely to withdraw and assume rigid postures in response to social stimulation (Rosenblum, 1961; Hansen, 1962). Similarly, in caged groups, males were found to initiate more play than females, and such play was more likely to be rough-and-tumble than was play initiated by females (Hinde & Spencer-Booth, 1967a).[4] If peer–peer play becomes violent, it may be interrupted by a mother, a male, or an aunt.

Interactions among immature monkeys involve a number of types of behavior in addition to those described. Infants show, for example, much incomplete sexual behavior. The development of such behavior into the adult sexual patterns has been described in detail by Harlow and Harlow (1965). The relation among the tentative sexual patterns of monkeys in their early months, the near-complete patterns shown by adolescents, and the complete adult patterns, does not seem to differ in kind from that between the various aggressive patterns seen at similar ages. The label "play" is equally suitable or unsuitable for the earlier stages. Although infants of both sexes may take either active or passive sexual roles, mounting is more characteristic of males and presenting of females (Hansen, 1962; Hinde & Spencer-Booth, 1967a).

Grooming is another adult activity that appears gradually in the infants, but in a group situation this is directed most towards the mother, less frequently towards aunts, and only rather infrequently towards other infants. Reciprocally, grooming of infants by other infants was much less frequent than grooming by the mother or by aunts (Hinde & Spencer-Booth, 1967a).

[4] Early hormonal influences in determining pre-puberal sex differences in a wide spectrum of the behavior of rhesus monkeys have been reviewed by Goy (1968).

Investigatory and maternal behavior is frequently shown towards infants by other infants. In the Madingley groups, this was much more common in females than in males. Among the former, it was more common in animals 2 years older than the infant in question than in females only 1 year older or more than 2 years older (see Section IV, B).

2. OTHER SPECIES

Data on the play behavior of other species is mostly of a qualitative nature. The nature of the most frequently used play patterns differ somewhat among species in relation to the habitat and adult repertoire. For instance, fast running and other locomotor activities are important in addition to play-biting and wrestling in the ground-living patas monkey (Hall, 1965). In tree-living species, climbing and swinging may predominate (e.g., Carpenter, 1934; Sugiyama, 1965a, 1965b). In many species, much of the play involves groups of six or more infants (e.g., patas monkeys, Hall, 1965; langur, Jay, 1965; and many others). Laboratory studies undoubtedly tend to underestimate the complexity of play patterns that may occur in nature. Wild-living Campbell's mona monkeys have been seen to play "King-of-the-castle", and to chase hornbills away from their roosting sites (Bourlière, et al., 1969; see also Bertrand, 1969, for data on the stumptailed macaque, Macaca arctoides). Play in apes is even more complex. Gorillas include "King-of-the-mountain" and "Follow-the-leader" (Schaller, 1963), and in chimpanzees, play involves a great diversity of patterns, including tickling and poking as well as aggressive patterns. Chimpanzees, unlike most if not all monkeys (but like some carnivores), often use objects, such as twigs or fruit, in their play (Loizos, 1967; van Lawick-Goodall, 1967).

There seem to be considerable interspecies differences in the age distribution of play, although some of these may stem from differences between field and laboratory conditions. In general, social play starts some time after the end of the first month in monkeys (feral baboons, fourth to sixth months, DeVore, 1965; captive baboons, fourth weeks, Rowell, Din, & Omar, 1968; feral langurs, about 3 months old, becoming frequent during fifth and sixth months, Sugiyama, 1965a; feral mangabeys, ninth to twelfth weeks, Chalmers, 1967; captive bonnet and pigtailed macaques, second month, Kaufman & Rosenblum, 1969). Gorillas (Schaller, 1963; Fossey, personal communication) and chimpanzees (van Lawick-Goodall, 1968) show the beginnings of social play while still being held by their mothers at 3 to 4 months of age.

There are even greater interspecies differences in the age at which play ceases. In feral rhesus, for instance, play is most frequent in juveniles (1 to 4 years old), but was not recorded in adults by Southwick, Beg, and Siddiqi (1965); and among captive rhesus adults, it occurs only in occasional individuals. Likewise, in his study of patas monkeys, Hall (1965) had only one brief record of an adult male playing. In the bonnet macaque, by contrast, adult males play regularly and often (Simonds, 1965), and they sometimes do so in other species (e.g., Bernstein, 1964). Squirrel monkeys still play quite a lot when 30 months of age. In chimpanzees, the frequency of play declines with age, but adults nevertheless show quite a considerable amount of play behavior (Fig. 11; van Lawick-Goodall, 1969). Mature gorillas also play fairly frequently, although social play

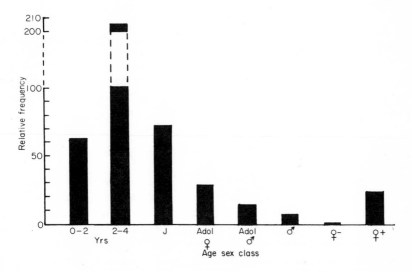

FIG. 11. Frequency of play sessions in different age/sex categories of chimpanzees. Ordinate gives relative frequency of play sessions. (Play sessions per 100 observation periods) J = Juvenile, A = Adolescent, ♀ – =Mature females without offspring, ♀ + = Mature female with offspring. (After van Lawick-Goodall, 1968)

is initiated more often by males than by females (Fossey, personal communication). Carpenter (1934) suggests that the cessation of play behavior is a consequence of pain and frustration, which occurs as aggressive play becomes more vigorous and more like real fighting. On this basis, it might be expected to be less common in the adults of more aggressive species, but it seems doubtful if this is the case (see Chapter 2 by Bernstein).

In a number of species, males play more vigorously than females, and females stop playing earlier than males; this follows the formation of play groups of uniform age and sex (e.g., Sugiyama, 1965a, b; Kummer, 1968b). Primarily on the basis of observations on cynomolgus monkeys (*Macaca fascicularis*). Fady (1969) listed a number of factors that appear to govern the selection of a play companion:

(a) Young monkeys tend to play within their own age group.

(b) Infants of high-ranking mothers play more frequently among themselves, while infants of low-rank mothers exhibit no preference related to social rank.

(c) Siblings are played with more than other young of comparable age.

(d) Young monkeys tend to play with others from the same "maternal club", that is, with the infants of mothers who habitually associated with their own.

The experimental data on rhesus monkeys cited earlier demonstrate the importance of peer–peer interaction for normal behavioral development. Evidence that the precise nature of that interaction could markedly influence adult behavioral characteristics is provided in Baldwin's (1969) study of squirrel monkeys. Infants born early in the birth season at first have few playmates other than the stronger and behaviorally more sophisticated juveniles, while infants born late are the weakest and most helpless of their age class.

D. Males

Of all aspects of the social environment, the role of the male seems to vary most between species. At one extreme lie the marmosets and tamarins where, in many species, the infant is carried more by the male than by the female (e.g., Hill, 1957; Hampton, Hampton, & Landwehr, 1966). This seems also to be the case with the night or owl monkey (*Aotus trivirgatus*) (Moynihan, 1964) and with the titi (*Callicebus moloch*) (Mason, cited by Napier & Napier, 1967). In the mangabey, the infants are often carried by adult males (Chalmers, 1967, 1968). Such a sharing of the burden between the sexes may well be adaptive in species with well-established pair bonds or where the social structure is derived from a single male–single female family. However, it is not yet certain that this is so in all the cases cited above.

At the other extreme are species like the patas monkey, where the male normally stays somewhat apart from the females in his troop (Hall, 1965), and even in captivity rarely shows interest in infants. If he does threaten one, mother and aunts may jointly threaten him back (Hall & Mayer, 1966, 1967; Hall, 1967).

Most species lie between these two extremes. The males protect the infants from predators and intruders, show some interest in and considerable tolerance of them, and sometimes interfere in disputes protectively on their behalf, but rarely show anything approaching maternal behavior. The langur, for instance, lies towards the patas end of the scale, though the extent of its paternal behavior may vary locally. Jay (1963) found that the male rarely approaches the newborns, and shows no response to them, and Sugiyama (1965b, 1967) records that when male langurs take over a troop they may kill all the infants in it. The young killed in this way are apparently always the offspring of a previous leader, and not their own. However, Sugiyama also records that adult males may allow infants and juveniles to embrace him, hang from his tail and play-wrestle with him (see also Yoshiba, 1968). In some areas, at least, the older juvenile langurs interact in a rather surprising way with the adult males. At first they merely approach, running up and veering away, but later they may even mount the male, or embrace him ventro–ventrally. This behavior may continue until the male juvenile is 4 years old, but does not occur in females. In addition, an infant may slap at a male copulating with its mother (Jay, 1963). In a different study area Yoshiba (1968) recorded that infants often approached and played with the adult male, but were never seen to mount him.

In the rhesus macaque, the infants soon learn to avoid adult males, though the latter are surprisingly gentle and tolerant of the infants' play. Wild adult males rarely initiate interaction with infants (Southwick, Beg, & Siddiqi, 1965; Kaufmann, 1966), though 2 out of 10 castrates released into a wild population showed interest in infants (Wilson & Vessey, 1968). In captive groups only occasional individual adult males show any positive interest in infants. However, one male in the Madingley colony played not infrequently with infants, and others were even seen to cuddle and carry in the ventro–ventral position infants whose mothers had been temporarily removed from the group (Spencer-Booth & Hinde, 1967). In these captive groups younger males showed rather more interest in infants than did adults. Thus, a higher proportion of males about 2 years older than the infants interacted with them than did older or younger males (Spencer-Booth, 1968a). In captive groups of pigtailed macaques, adult males showed little tendency to interact with infants, but bonnet males were solicitous towards maternally deprived infants, and sometimes cradled and carried them (Kaufman & Rosenblum, 1969).

In some other macaques, the adult males show even more tendency to interact with infants. In the stumptailed macaque, Bertrand (1969) records a 2-year-old male acting protectively to a 4-month-old male, and describes a special relationship between a fully adult male and an infant. In the Japanese macaque, on the whole, infants avoid the adult males, but particular male–infant relationships sometimes are established, the male often carrying the infant. Although males of high social-rank are sometimes involved, it is especially common in middle-rank males of caste-structured troops, and it may help the male to become established in the troop. In the Barbary ape, where considerable paternal behavior has been recorded in captivity (Lahiri & Southwick, 1966), field observations by Deag and Crook (personal communication) reveal relations of special interest between males and infants. As noted above, in this species the mothers are permissive, the infants ranging widely and interacting often with other animals in the group. The males show a marked tendency to interact with young babies. They may carry babies when the group is in a dangerous situation, and play with them when the group is relaxed. This play is almost entirely restricted to juvenile and infant males. Often the carrying of a baby by a male develops out of a grooming session; a male joins a female and her baby to form a grooming trio, and later leaves with the baby. Sometimes males seem to use babies as a "passport" for approaching another animal in situations where tension is already or is potentially present. In such cases, one male with a baby, usually on its back, approaches another male and either sidles up to it in a manner somewhat resembling sexual presentation, or sits down and puts the baby between them. The approaching male, who is usually the subordinate of the two, often seems to seek out a particular individual to approach. His possession of a baby increases the probability that he will stay near the dominant animal. Although such observations indicate that the male–infant interaction results from male initiative, the reverse is occasionally the case.

In the baboons, also, quite a considerable amount of paternal behavior may be shown. Field observations by Hall and DeVore (1965) and DeVore (1963) suggest an age distribution of paternal responsiveness in chacma and olive baboons somewhat different from that in the rhesus. Juvenile and subadult males show only perfunctory interest in infants, but some adult males show considerable involvement, sometimes carrying and often protecting them. They are tolerant of black infants and protect them from older infants during their play. Their interest in the babies reaches its peak when the infants are 2.5 to 4 months old, but they remain protective of older ones, and may form special relationships with them. (Rowell, personal communication, found a rather different time course; adult males showed interest in infants from about 5 to 13 weeks, juvenile males slightly later.) Bolwig (1959) recorded the adoption of two chacma baboon infants by males, though most males were intolerant of infants. The most interesting data here come from a study of olive baboons at the Gombe Stream National Park by Ransom and Ransom (personal communication). They record male–infant relationships arising in several ways:

1. A relationship between a mother and a male, perhaps existing before the birth of the infant, fosters a relationship between the male and the infant.

2. A mother continuously rejects her infant and a high-dominant male temporarily adopts it, staying with it and protecting it over prolonged and recurring periods of time.

3. A male forms a relationship with a particular infant which is accompanied by mere mutual tolerance between the male and the mother, and no special rejection of the infant by the mother.

4. Two particular high-dominant males paid frequent attention to the infants of high-dominant females.

5. A male's possession of an infant may temporarily enhance his dominance status, or at least ameliorate an agonistic situation. This may arise from one of the previously-mentioned types of relationship, or independently.

In hamadryas baboons male–infant relationships develop through an onto-genetic sequence that provides the raw material for troop organization (Kummer, 1967, 1968a, b). The troops in this species consist of several stable bands, each in turn consisting of one-male units comprising a male and one or more females, sometimes accompanied by subadult males. The subadult males sometimes kidnap temporarily, or may even adopt, infants. In addition subadult or young adult males often form the center of play groups. The juveniles will run to the male and clasp him if frightened, and he may then threaten the aggressor. The male thus temporarily takes the role of the protective mother for each infant in turn. Since the infant, by turning round and screaming at the aggressor, is able to induce the male to threaten the latter, it is able to establish a triangular relationship with the male and the aggressor, the male being in the maternal role. Since the male sometimes takes the protective role on behalf of other infants, however, he becomes an ambivalent figure, protective to the infant nearest to him but aggressive to infants farther away. This in turn means that the infants compete for proximity to the male. The males come to adopt particular infants for progressively longer periods. This may ameliorate the process of weaning for the infants. Soon after the females are 1 year old, they enter into a consort relationship with a young adult male. The male then takes on the maternal role towards the female, but also ensures her continued proximity by attacking her if she strays. These groups develop into the normal groups of one male and one or more females, in which the competition between the females for proximity to the male leads to a constant matrix of interactions. The threat of a female towards a rival not only induces the male also to threaten but, since she threatens from nearby the male, makes it almost impossible for the other animal to threaten back, since the threat would be directed towards the male. Thus the adult male–female bond seems to arise from a transferred mother–child relationship.

Subadult males also employ the maternal role when they are themselves involved in agonistic situations. Typically, a frightened subadult male grasps an infant and turns away from the adult male. Alternatively, he may invite an infant or juvenile to jump on his back. As in other species, such behavior may reduce the probability that the dominant male will attack. It is also possible that the "contact comfort" reduces the stress in the subadult male. This tendency of hamadryas baboon males to take a maternal role in agonistic situations is in harmony with their tendency to behave maternally toward infants. The fully adult male remains tolerant to young infants and will occasionally hold them, but such behavior is sporadic.

Adult male gorillas are protective and tolerant of infants, but only rarely carry them (Schaller, 1963). Fossey (personal communication) records that they often tolerate infants playing on or near their bodies, and sometimes take infants from their mothers to examine or groom them, though they do not treat the infants as a female would. Consort relationships between silverbacks and older infants are common. A study by van Lawick-Goodall (1968) records juvenile male chimpanzees in the Gombe Stream National Park, but not adult males, showing protective behavior to, and carrying, younger siblings. However, Albrecht (personal communication), working near Kankasali (Republic of Guinea), has seen an adult male chimpanzee apparently looking after three infants, with no other chimpanzees close by. This male gave food to the infants when they begged.

It is apparent that the nature and extent of male–infant relations vary widely among species (see also references in Mason, 1965) and even among individuals within any one species. The bases for these variations in either functional (e.g., in relation to the social structure or environment) or causal terms are not yet understood, but it is perhaps worth stressing that in the rhesus macaque at least, the sex difference in frequency of "maternal" responsiveness appears even in adolescent animals (e.g., Spencer-Booth, 1968a). This suggests that maternal behavior is not necessarily dependent upon a contemporary hormonal state associated with parturition or lactation.

V. CONCLUSION: THE SOCIAL NEXUS

A. Introduction

In the previous sections, the interactions among young primates and various types of social companions have been discussed. These interactions occur within a social group whose structure varies markedly among and even within species (see, e.g., DeVore, 1965; Jay, 1965; Crook, in press; see also Chapter 2 by Bernstein). In this section, some further aspects of the diversity of the social milieu in which the infant develops will be considered.

B. The Diversity of Interactions with Each Type of Social Companion

Each type of social companion interacts with the infant in diverse ways. The mother, for instance, provides not only food and warmth, but "contact comfort," physical surfaces for tactile exploration and for the practice of visuomotor coordination, support and protection; she grooms the infant and is groomed by it; she may act as a play companion, provide stimuli that elicit its aggressive, sexual, and fleeing behavior, and so on through a list that could include virtually all the types of social interaction in which monkeys indulge. Except for providing milk, each other type of social companion may also interact with the infant in nearly every one of these ways, though the relative extent to which they do so varies with their age–sex category. Other infants, for instance, may groom, play, elicit sexual, aggressive, and fleeing behavior, and provide practice in visuomotor coordination. Furthermore, in the absence of the mother, they may provide contact comfort and warmth (e.g., Harlow & Harlow,

1965, 1969). With few exceptions (such as "contact comfort"), we cannot manipulate these facets of the social interactions of the young animal independently. We therefore classify the social influences in terms of the individual who exerts them—mother, peer, and so on. Yet it may be that to the infant monkey, the total amount of each type of interaction in which it can engage has an importance that is independent of the type of individual with which it does so: perhaps, for instance, an infant may be little affected by the absence of peers if its mother plays with it. Experiments in which the development of infant monkeys was studied in the absence of one or more categories of social companion, while revealing many aspects of the dynamics of development (e.g., Harlow & Harlow, 1965), have given rise to controversy concerning the extent to which the several categories are essential for normal development (see the review by Jensen & Bobbitt, 1967). Recent experiments by Bauer, Alexander, and Rosenblum (reviewed by Seay, 1966) indicate that the effects of inadequate maternal care in rhesus monkeys can be ameliorated by peer experience, and vice versa, a finding clearly to be interpreted in terms of the multiplicity of types of interaction possible between the infant and each category of social companion.

However, such an approach could lead us to oversimplify if we supposed that a given amount of one type of interaction (e.g., playing) with one type of social companion (e.g., peers) was *always* equivalent to a similar amount with another (e.g., the mother). This is not necessarily the case. For example, the development of a monkey with two social companions—one that attacked it often and one that avoided it—might well be different from that occurring if one of the two companions was unresponsive to it and the other alternately attacked and avoided it to an amount equal to that in the first case. In the human case, development normally involves a focusing of the attachment behavior of the infant onto one individual (usually the mother), and this "monotropy" may be important for its mental health (Bowlby, 1969). One study of the effects of rotating mothers among rhesus infants was cited in Section IV, A, 10.

C. External Influences on the Nature of Interactions

In a number of ways, interaction between the infant and any other individual may be affected by factors external to both of them. This is shown most clearly by some of the Wisconsin social deprivation experiments. For instance, infants raised without mothers but in small groups spend much of their time clasping each other, apparently deriving from each other some of the "contact comfort" that they would normally have obtained from their mothers. Furthermore, the richness of the physical environment may affect the nature of the mother–infant relationship. Jensen, Bobbitt, and Gordon (1967) have shown that environmental privation may retard the increasing independence of infant from mother (see Section IV, B, 7).

In a social situation, such effects are of course even more complex. Thus, considering the four categories of social companion discussed earlier, the mother influences the interactions of the infant with other infants, males, and aunts through her restrictiveness–permissiveness with the infant and her dominance–subordinance relations with the others. The infant's presence may reciprocally affect the mother's relations with her social companions, causing her to withdraw from their immediate proximity, especially at the time of the birth, transforming her into a center of interest, causing her to attack or threaten them in its defense, and giving her protection and status in aggressive interactions. Both the aunts and peers (Harlow & Harlow, 1969) may affect the infant's relations with its mother and with each other. Males may affect the infant's and mother's interactions with all other categories of social companions (see, e.g., Stynes, Rosenblum, & Kaufman, 1968), and in some species the infant's presence or proximity may affect the males' social relations. Quantitative data on such effects in captive rhesus groups of different composition are given by Bernstein & Draper (1964). Clearly, the possibilities are innumerable, and we are a very long way from understanding the nexus of social forces that surround the developing infant.

D. Tripartite Relations

So far, we have considered two levels of complexity—direct interactions between the infant and its social companions, and the ways in which these interactions may be influenced by third parties. A further level concerns tripartite relations, that is, those in which, according to Kummer (1967), "three individuals *simultaneously* interact in three *essentially different roles* and *each of them aims its behavior at both* of its partners [p. 64]." Such a situation arises, for example, with a protector, a protected and an antagonist, as for instance when an infant approached by another individual retires to its mother, who threatens the intruder. As we have seen, in the hamadryas baboon, this situation recurs throughout development, but the role of protective mother is transferred to a male. The infant, having run to the male, may turn around and scream at the intruder. The infant seems to learn that sitting close to the male and screaming at the intruder is likely to induce aggression from the former and submission from the latter. The relationship is complicated because several individuals transfer the mother-role to the same male, and therefore compete with each other for his proximity. The same situation recurs with the females in the one-male groups, and Kummer (1967, 1968a, b) argues convincingly that this transfer of roles from the mother–infant–aggressor situation plays an essential role in their formation (see also Bowden, 1966). A somewhat similar process appears to be involved in the somewhat looser social organization of the Japanese macaque (Itani, 1963).

E. Roles within the Primate Group

As the preceding discussion shows, various types of response to the infant are shown by the several age–sex classes to different extents. Thus it is possible to specify spectra of interaction probabilities for each age–sex class showing that, for instance, the probability of adult females showing certain responses to the infants is different from that of adult males. One way of describing this is to say that the different age–sex classes have different roles with respect to the infant. This is a particular aspect of their differing roles in the structure of the society, for many aspects of the behavior of individual animals are predictable in terms of their age, sex and parental status (e.g., Bernstein & Sharpe, 1966). The usefulness of specifying the roles of the animals constituting primate groups has recently been emphasized by Gartlan (1968). Studying vervets, he was able to categorize all social interactions (omitting maternal and play) as territorial display, social vigilance, social focus, friendly approach, territorial chasing, punishing, and leading. He then constructed a matrix showing the contribution of each age–sex class (or individual) to each "role-category." Such a procedure is potentially capable of showing that social roles are not distributed at random through the population. The identification in this way of the types of behavior shown by each age–sex class will facilitate comparisons of social organization both within and among species. Gartlan's paper marks an important step forward to this end. As discussed elsewhere (Hinde, in press), however, the use of the "role" concept in studies of nonhuman primates has run into both semantic and statistical problems that must be solved before much further progress is possible. In particular, the extent to which it carries implications of adaptedness must be made explicit.

F. Individual Relationships

In the previous sections, interactions between infant primates and their social companions have been described in terms of age–sex categories. This is a convenient way of classifying the data, but it is not to be forgotten that interactions occur between individuals. Furthermore, particular relationships may be formed between individuals, which transcend the particular interactions from which they arise. Two types of such relationships must be mentioned.

1. *Dominance relations.* "Dominance," as usually used, implies that one individual has precedence over another in a variety of situations, and causes it to retreat in agonistic encounters. In small groups of captive primates, the animals can often be arranged in a "dominance hierarchy" which permits prediction of a wide variety of social interactions. Under natural conditions, likewise, relationships of this type between pairs of individuals may be fairly stable and permit prediction of access to females (e.g., baboons, DeVore, 1965) or to food, or right of way (e.g., chimpanzees, van Lawick-Goodall, 1968).

However, the old view that a rigid hierarchical structure imposed from above is essential for the integration of groups requires modification in a number of respects (e.g., Rowell, 1966; Gartlan, 1968). In the first place, the concept of dominance requires careful handling. Its status as an explanatory device must be constantly borne in mind, and circular argument of the type: "Dominance is assessed by priority of access; an animal has priority of access because it is dominant" avoided (Gartlan, 1968). To turn to more empirical issues, in many species dominance interactions are inconspicuous or absent in the field (e.g., Carpenter, 1942; Bernstein & Schusterman, 1964), though they may be accentuated in captivity (e.g., patas monkeys, Hall, Boelkins, & Goswell, 1965). Where they are present, dominance relations are seldom so simple or so rigid as the classic "dominance hierarchy" concept implies (Rowell, 1966; Gartlan, 1968). Indeed, the social relations are determined through learning by both participants in, and observers of, every social encounter (Hall, Boelkins, & Goswell, 1965; Rowell, 1966). Dominance relations found in one context (e.g., over food) may not apply in others (e.g., sex behavior). Among hamadryas baboon females, there seem to be two ranking orders, an autonomous one among the females, and another based on male preferences. Amongst the males, displacement of one animal from an incentive by another is extremely rare, and dominance orders based on the number of females each male possesses or on his influence on travel direction do not coincide (Kummer, 1968b). Furthermore, in many species the order is determined as much by the behavior of the subordinate animal as by the dominant one. From observations on captive baboons Rowell (1966) stresses that "it is the subordinate animal which cautiously observes and maintains a hierarchy, while a dominant animal could almost be defined as one which does not 'think before it acts' in social situations [p. 441].

It must also be noted that most social encounters do not involve merely two individuals, but are influenced by the proximity of, or participation by, others. Baboon troops are dominated by a small group of males who depend upon one another for mutual support. A member of this group may be subordinate to other troop members when alone (DeVore, 1965). Similarly, within a group of high-ranking male chimpanzees, the outcome of an encounter between two individuals was affected by which others were present (van Lawick-Goodall, 1968). Again, it is well established that interactions between an infant and its peers or other adults may be affected by the presence of its mother. It is thus not surprising that dominance orders as assessed in extragroup encounters between two individuals may have little relevance to those in the group (Maslow, cited by Altmann, 1967).

With these and other reservations (see Gartlan, 1968; Rioch, 1967), it remains true that the outcome of agonistic interactions and precedence to objects or situations between individuals remain stable at least over limited periods (e.g.,

Kaufmann, 1967). As we have seen, such dominance relations may be important in interactions between mothers and aunts, between peers, and in many other contexts.

2. *Affectional bonds.* Some relationships merit description in terms of affectional bonds between individuals. The mother–infant relationship is, as we have seen, perhaps the strongest of all such relationships; and in both natural and captive groups it may persist long after the period of juvenile dependence (e.g., Sade, 1965; van Lawick-Goodall, 1968). Comparable bonds may be formed between siblings and, under laboratory conditions at least, between peers brought up in the same cage (e.g., Harlow, 1969). In some species, at least, quite strong affectional bonds may develop between unrelated individuals (e.g., Kummer, 1968a,b). That the formation of an affectional bond with a particular individual in early life is essential for subsequent normal social relations, as has been claimed for our own species (Bowlby, 1969), seems probable. Both rhesus monkeys and chimpanzees reared in early social isolation fail to form affectional bonds to peers later on (Mason, 1964; and Section III), and there is a good deal of evidence from field studies that the one type of bond arises from the other. Such bonds may be of crucial importance not only in the determination of social relationships (e.g., Harlow, 1969), but also in a number of other contexts. For instance "cooperative"[5] behavior, such as that described by Harlow (1969) in rhesus monkeys which caught alligators from the moat surrounding their enclosure, and by van Lawick-Goodall (1968) for chimpanzees, would seem to depend on bonds of this sort.

G. Conclusion

That the behavioral development of nonhuman primates can be markedly affected by severe social deprivation during the first year of life is proven. That the more subtle differences in the social environment that occur between individual infants under natural conditions can produce differences between their behavioral characteristics is not yet proven, but seems probable on the currently available, admittedly circumstantial, evidence. Effects are likely to be produced not only through the relationships with the mother and the peers, but through relationships with other animals, which may affect the emotional development of the infant and may also influence what it learns from its environment. That the social environment is extremely complex, and differs markedly between species, will be apparent from the preceding discussion (see also Chapter 2 by Bernstein). The interspecies differences in the social environment of the young are reflections of differences in social structure.

---

[5] Dr. van Lawick-Goodall has pointed out to me the wisdom of using the term "cooperative" in a descriptive sense to refer only to the result of the behavior.

Whether the effects are circular, the differences in social structure being in part a consequence of the social environment in which the young develop, is a matter worth further study (e.g., Crook, 1970).

ACKNOWLEDGMENT

I am grateful to Drs. Jane van Lawick-Goodall, Thelma Rowell, G. P. Sackett and Yvette Spencer-Booth for their comments on earlier drafts of the manuscript, and to various authors mentioned in the text for permitting me to cite unpublished work.

REFERENCES

Ainsworth, M. D. The effects of maternal deprivation: A review of findings and controversy in the context of research strategy. *In Public Health papers No. 14.* Geneva: World Health Organization, 1962. Pp. 97-165.

Altmann, S. A. Field observations on a howling monkey society. *Journal of Mammalogy,* 1959, **40,** 317-330.

Altmann, S. A. Discussion of reproductive behavior. *In* S. A. Altmann (ed.), *Social communication among primates.* Chicago: Univ. Chicago Press, 1967. Pp. 55-59.

Baldwin, J. D. The ontogeny of social behavior of squirrel monkeys (*Saimiri sciureus*) in a semi-natural environment. *Folia Primatologica,* 1969, **11,** 35-79.

Bell, R. Q. A reinterpretation of the direction of effects in studies of socialization. *Psychological Review,* 1968, **75,** 81-95.

Bernstein, I. S., & Draper, W. A. The behaviour of juvenile rhesus monkeys in groups. *Animal Behaviour,* 1964, **12,** 84-91.

Bernstein, I. S., & Sharpe, L. G. Social roles in a rhesus monkey group. *Behaviour,* 1966, **26,** 91-104.

Bernstein, I. S., & Schusterman, R. J. The activities of gibbons in a social group. *Folia Primatologica,* 1964, **2,** 161-170.

Bertrand, M. *The behavioral repertoire of the stumptail macaque.* Basel: Karger, 1969. (*Bibliotheca Primatologica,* No. 11).

Bolwig, N. A study of the behaviour of the chacma baboon *Papio ursinus. Behaviour,* 1959, **14,** 136-163.

Bolwig, N. Bringing up a young monkey. *Behaviour,* 1963, **21,** 300-330.

Booth, A. H. Observations on the natural history of the olive colobus monkey (*Procolobus verus). Proceedings of the Zoological Society of London,* 1957, **129,** 421-430.

Booth, C. Some observations on behavior of *Cercopithecus* monkeys. *Annals of the New York Academy of Sciences,* 1962, **102,** 477-487.

Bourlière, F., Bertrand, M., & Hunkeler, C. L'Ecologie de la Mone de Lowe. *La Terre et la Vie,* 1969, **2,** 135-163.

Bowden, D. Primate behavioral research in the USSR. *Folia Primatologica,* 1966, **4,** 346-360.

Bowlby, J. *Maternal care and mental health.* London: World Health Organization, 1951.

Bowlby, J. *Attachment and loss.* Vol. I. *Attachment.* London: Hogarth Press, 1969.

Carpenter, C. R. A field study of the behavior and social relations of howling monkeys (*Alouatta palliata*). *Comparative Psychology Monographs,* 1934, **10,** No. 2 (Whole No. 48).

Carpenter, C. R. Societies of monkeys and apes. *Biological Symposia. 1942,* **8,** 177-204.

Carpenter, C. R. The howlers of Barro Colorado Island. *In* I. DeVore (ed.), *Primate Behavior.* New York: Holt, Rinehart & Winston, 1965. Pp. 250-291.

Chalmers, N. R. The ethology and social organization of the black mangabey (*Cercocebus albigena*). Unpublished doctoral dissertation, Cambridge Univ., 1967.

Chalmers, N. R. The social behavior of free living mangabeys in Uganda. *Folia Primatologica,* 1968, 8, 263-281.

Child, I. L. An experimental investigation of "taboo" formation in a group of monkeys. *Psychological Bulletin,* 1938, **35,** 705.

Crook, J. H. The socio-ecology of primates. *In* J. H. Crook (ed.), *Social behaviour in birds and mammals.* London: Academic Press, 1970. Pp. 103-166.

DeVore, I. Mother-infant relations in free-ranging baboons. *In* Harriet L. Rheingold (ed.), *Maternal behavior in mammals.* New York: Wiley, 1963. Pp. 305-335.

DeVore, I. Male dominance and mating behavior in baboons. *In* F. A. Beach (ed.), *Sex and behavior.* New York: Wiley, 1963. Pp. 305-335.

Epple, G. Verleichende Untersuchungen über Sexual- und Sozialverhalten der Krallenaffen (Hapalidae). *Folia Primatologica,* 1967, 7, 37-65.

Evans, C. S. Methods of rearing and social interaction in *Macaca nemestrina. Animal Behaviour,* 1967, **15,** 263-266.

Fady, J. C. Les jeux sociant: le compagnon de jeux chex les jeunes. Observations chez *Macaca irus. Folia Primatologica,* 1969, **11,** 134-143.

Frisch, J. E. Individual behavior and intertroop variability in Japanese macaques. *In* Phyllis C. Jay (ed.), *Primates.* New York: Holt, Rinehart & Winston, 1968. Pp. 243-252.

Fuller, J. L. Experimental deprivation and later behavior. *Science,* 1967, **158,** 1645-1652.

Gardner, R. A., & Gardner, Beatrice T. Teaching sign language to a chimpanzee. *Science,* 1969, **165,** 664-672.

Gardner, Beatrice T., & Gardner, R. A. Two-way communication with an infant chimpanzee. *In* A. M. Schrier & F. Stollnitz (eds), *Behavior of nonhuman primates.* Vol. 4. New York and London: Academic Press, 1971.

Gartlan, J. S. Structure and function in primate society. *Folia Primatologica,* 1968, 8, 89-120.

Gartlan, J. S., & Brain, C. K. Ecology and social variability in *Cercopithecus aethiops* and *C. mitis. In* Phyllis C. Jay (ed.), *Primates.* New York: Holt, Rinehart & Winston, 1968. Pp. 253-292.

Goodall, Jane. Tool-using and aimed throwing in a community of free-living chimpanzees. *Nature,* 1964, **201,** 1264-1266.

Goodall, Jane. Chimpanzees of the Gombe Stream reserve. *In* I. DeVore (ed.), *Primate behavior.* New York: Holt, Rinehart & Winston, 1965. Pp. 425-473.

Goswell, M. J., & Gartlan, J. S. Pregnancy, birth and early infant behaviour in the captive patas monkey *Erythrocebus patas. Folia Primatologica,* 1965, 3, 189-200.

Goy, R. W. Organizing effects of androgen on the behaviour of rhesus monkeys. *In* R. Michael (ed.), *Endocrinology and human behaviour.* London: Oxford Univ. Press, 1968. Pp. 12-31.

Griffin, G. A., & Harlow, H. F. Effects of 3 months of total social deprivation on social adjustment and learning in the rhesus monkey. *Child Development,* 1966, 37, 533-547.

Hall, K. R. L. Numerical data, maintenance activities and locomotion of the wild chacma baboon, *Papio ursinus. Proceedings of the Zoological Society of London,* 1962, **139,** 181-220.

Hall, K. R. L. Behaviour and ecology of the wild patas monkey, *Erythrocebus patas,* in Uganda. *Journal of Zoology,* 1965, **148,** 15-87.

Hall, K. R. L. Social interactions of the adult male and adult females of a patas monkey group. *In* S. A. Altmann (ed.), *Social communication among primates.* Chicago: Univ. Chicago Press, 1967. Pp. 261-280.

Hall, K. R. L. Social learning in monkeys. *In* Phyllis C. Jay (ed.), *Primates.* New York: Holt, Rinehart & Winston, 1968. Pp. 383-397.

Hall, K. R. L., & DeVore, I. Baboon social behavior. *In* I. DeVore (ed.), *Primate behavior.* New York: Holt, Rinehart & Winston, 1965. Pp. 53-110.

Hall, K. R. L., & Mayer, B. Social interactions in a group of captive patas monkeys, *Erythrocebus patas. Folia Primatologica,* 1966/1967, **5,** 213-236.

Hall, K. R. L., Boelkins, R. C., & Goswell, M. J. Behaviour of patas monkeys, *Erythrocebus patas,* in captivity, with notes on the natural habitat. *Folia Primatologica,* 1965, **3,** 22-49.

Hamburg, D. A. Observations of mother-infant interactions in primate field studies. *In* B. M. Foss (ed.), *Determinants of infant behaviour.* Vol. IV. London: Methuen, 1969. Pp. 3-14.

Hampton, J. K., Hampton, Suzanne H., & Landwehr, B. T. Observations on a successful breeding colony of the marmoset, *Oedipomidas oedipus. Folia Primatologica,* 1966, **4,** 265-287.

Hansen, E. W. *The development of maternal and infant behavior in the rhesus monkey.* (Doctoral dissertation, Univ. of Winconsin). Ann Arbor, Mich: University Microfilms, 1962. No. 63-653.

Hansen, E. W. The development of maternal and infant behavior in the rhesus monkey. *Behaviour,* 1966, **27,** 107-149.

Hansen, E. W., & Mason, W. A. Socially mediated changes in lever-responding of rhesus monkeys. *Psychological Reports,* 1962, **11,** 647-654.

Harlow, H. F. Age-mate or peer affectional system. *In* D. S. Lehrman, R. A. Hinde, & Evelyn Shaw (eds), *Advances in the study of behavior.* Vol. 2. New York: Academic Press, 1969. Pp. 333-383.

Harlow, H. F., & Harlow, Margaret K. The affectional systems. *In* A. M. Schrier, H. F. Harlow, & F. Stollnitz (eds), *Behavior of nonhuman primates.* Vol. II. New York: Academic Press, 1965. Pp. 287-334.

Harlow, H. F., & Harlow, Margaret K. Effects of various mother-infant relationships on rhesus monkey behaviors. *In* B. M. Foss (ed.), *Determinants of infant behaviour.* Vol. IV. London: Methuen, 1969. Pp. 15-36.

Harlow, H. F., Schiltz, K. A., & Harlow, Margaret K. Effects of social isolation on the learning performance of rhesus monkeys. *In* C. R. Carpenter (ed.), *Proceedings of the Second International Congress of Primatology.* Vol. I. *Behavior..* Basel: Karger, 1969. Pp. 178-185.

Hayes, K. J., & Hayes, Catherine. Imitation in a home-raised chimpanzee. *Journal of Comparative and Physiological Psychology,* 1952, **45,** 450-459.

Hill, W. C. Osman. *Primates. Comparative anatomy and taxonomy.* Vol. III. *Pithecoidea: Platyrrhini: Family Hapalidae.* Edinburgh: Edinburgh Univ. Press, 1957.

Hill, W. C. Osman. Laboratory breeding, behavioural development and relations of the talapoin (*Miopithecus talapoin*). *Mammalia,* 1966, **30,** 353-370.

Hinde, R. A. Analysing the roles of the partners in a behavioural interaction—mother—infant relations in rhesus macaques. *Annals of the New York Academy of Sciences,* 1969, **159,** 651-667.

Hinde, R. A. *Animal behaviour* (2nd ed.) New York: McGraw-Hill, 1970.

Hinde, R. A. Problems in the development of social behavior. *In* Ethel Tobach, L. R. Aronson, & Evelyn Shaw (eds), *The biopsychology of development.* New York: Academic Press, 1971. In press.

Hinde, R. A., & Atkinson, Sue. Assessing the roles of social partners in maintaining mutual proximity, as exemplified by mother—infant relations in rhesus monkeys. *Animal Behaviour,* 1970, **18,** 169-176.

Hinde, R. A., & Rowell, T. E. Communication by postures and facial expressions in the rhesus monkey (*Macaca mulatta*). *Proceedings of the Zoological Society of London,* 1962, **138,** 1-21.

Hinde, R. A., Rowell, T. E., & Spencer-Booth, Y. Behaviour of socially living rhesus monkeys in their first six months. *Proceedings of the Zoological Society of London,* 1964, **143,** 609-649.

Hinde, R. A., & Spencer-Booth, Y. The behaviour of socially living rhesus monkeys in their first two and a half years. *Animal Behaviour,* 1967, **15,** 169-196. (a)

Hinde, R. A., & Spencer-Booth, Y. The effect of social companions on mother—infant relations in rhesus monkeys. *In* D. Morris (ed.), *Primate ethology.* London: Weidenfeld & Nicolson, 1967. Pp. 267-286. (b)

Hinde, R. A., & Spencer-Booth, Y. The study of mother—infant interaction in captive group-living rhesus monkeys. *Proceedings of the Royal Society of London. Series B. Biological Sciences,* 1968, **169,** 177-201.

Hinde, R. A., & Spencer-Booth, Y. Towards understanding individual differences in rhesus mother—infant interaction. *Animal Behaviour,* 1971, **19,** 165-173.

Hinde, R. A., & Spencer-Booth, Y. Individual differences in the responses of rhesus monkeys to a period of separation from their mothers. *Journal of Child Psychology and Psychiatry and Allied Disciplines,* 1971. (In press.)

Hopf, Sigrid. Ontogeny of social behavior in the squirrel monkey. *In* D. Starck, R. Schneider, & H. J. Kuhn (eds), *Progress in primatology.* Stuttgart: Fischer, 1967. Pp. 255-262.

Imanishi, K. Social organization of sub-human primates in their natural habitat. *Current Anthropology,* 1960, **1,** 393-407.

Itani, J. On the acquisition and propagation of a new food habit in the natural group of the Japanese monkey at Takasaki Yama. *Primates,* 1958, **1,** 84-98.

Itani, J. Paternal care in the wild Japanese monkey *Macaca fuscata. Primates,* 1959, **2,** 61-94.

Itani, J. Paternal care in the wild Japanese monkey *Macaca fuscata. In* C. H. Southwick (ed.), *Primate social behavior.* Princeton: Van Nostrand, 1963.Pp. 91-97.

Jay, Phyllis C. Aspects of maternal behavior among langurs. *Annals of the New York Academy of Sciences,* 1962, **102,** 468-476.

Jay, Phyllis C. Mother—infant relations in langurs. *In* Harriet L. Rheingold (ed.), *Maternal behavior in mammals.* New York: Wiley, 1963. Pp. 282-304.

Jay, Phyllis C. The common langur of North India. *In* I. DeVore (ed.), *Primate behavior.* New York: Holt, Rinehart & Winston, 1965. Pp. 197-249.

Jensen, G. D. Reaction of monkey mothers to long-term separation from their infants. *Psychonomic Science,* 1968, **11,** 171-172.

Jensen, G. D., & Bobbitt, Ruth A. Implications of primate research for understanding infant development. *In* J. Hellmuth (ed.), *The exceptional infant. I.* Seattle: Special Child Publications, 1967.

Jensen, G. D., Bobbitt, Ruth, A. & Gordon, Betty N. The development of mutual independence in mother–infant pigtailed monkeys, *Macaca nemestrina. In* S. A. Altmann (ed.), *Social communication among primates.* Chicago: Univ. Chicago Press, 1967. Pp. 43-54.

Jensen, G. D., Bobbitt, Ruth A., & Gordon, Betty N. Sex differences in the development of independence of infant monkeys. *Behaviour,* 1968, **30,** 1-14. (a)

Jensen, G. D., Bobbitt, Ruth A., & Gordon, Betty N. Effects of environment on the relationship between mother and infant pigtailed monkeys *(Macaca nemestrina). Journal of Comparative and Physiological Psychology,* 1968, **66,** 259-263. (b)

Jensen, G. D., Bobbitt, Ruth A., & Gordon, Betty N. Studies of mother–infant interactions in monkeys *(Macaca nemestrina).* Hitting behavior. *In* C. R. Carpenter (ed.), *Proceedings of the Second International Congress of Primatology.* Vol. 1. *Behavior.* Basel: Karger, 1969. Pp. 186-193.

Jensen, G. D., Bobbitt, Ruth A., & Gordon, Betty N. Mothers' and infants' roles in the development of independence of *Macaca nemestrina. In* C. R. Carpenter (ed.), *Social regulatory mechanisms of primates.* (In press.)

Jensen, G. D., & Tolman, C. W. Mother–infant relationship in the monkey *Macaca nemestrina:* the effect of brief separation and mother–infant specificity. *Journal of Comparative and Physiological Psychology,* 1962, **55,** 131-136.

Joslin, J., Fletcher, H., & Emlen, J. A comparison of the responses to snakes of lab- and wild-reared rhesus monkeys. *Animal Behaviour,* 1964, **12,** 348-352.

Kaufman, I. C., & Rosenblum, L. A. The reaction to separation in infant monkeys: Anaclitic depression and conservation-withdrawal. *Psychosomatic Medicine,* 1967, **29,** 648-675.

Kaufman, I. C., & Rosenblum, L. A. The waning of the mother–infant bond in two species of macaque. *In* B. M. Foss (ed.), *Determinants of infant behaviour.* Vol. IV. London: Methuen, 1969. Pp. 41-60.

Kaufman, I. C., & Rosenblum, L. A. Effects of separation from mother on the emotional behavior of infant monkeys. *Annals of the New York Academy of Sciences,* **159,** 681-695.

Kaufmann, J. H. Social relations of infant rhesus monkeys and their mothers in a free ranging band. *Zoologica,* 1966, **51,** 17-28.

Kaufmann, J. H. Social relations of adult males in a free-ranging band of rhesus monkeys. *In* S. A. Altmann (ed.), *Social communication among primates.* Chicago: Univ. Chicago Press, 1967. Pp. 73-98.

Kawamura, S. The process of sub-culture propagation among Japanese macaques. *Primates,* 1959, **2,** 43-60.

Kellogg, W. N. Communication and language in the home-raised chimpanzee. *Science,* 1968, **162,** 423-427.

Koford, C. B. Group relations in an island colony of rhesus monkeys. *In* C. H. Southwick (ed.), *Primate social behavior.* Princeton: Van Nostrand, 1963. Pp. 136-152. (a)

Koford, C. B. Rank of mothers and sons in bands of rhesus monkeys. *Science,* 1963, **111,** 356-357. (b)

Koford, C. B. Population dynamics of rhesus monkeys on Cayo Santiago. *In* I. DeVore (ed.), *Primate behavior.* New York: Holt, Rinehart & Winston, 1965.

Kummer, H. Tripartite relations in hamadryas baboons. *In* S. A. Altmann (ed.), *Social communication among primates.* Chicago: Univ. Chicago Press, 1967. Pp. 63-72.

Kummer, H. Two variations in the social organization of baboons. *In* Phyllis C. Jay (ed.), *Primates.* New York: Holt, Rinehart & Winston, 1968. Pp. 293-312. (a)

Kummer, H. *Social organization of hamadryas baboons: A field study.* Basel: Karger, 1968. *(Bibliotheca Primatologica,* No. 6) (b)

Kummer, H., & Kurt, F. A comparison of social behavior in captive and wild hamadryas baboons. *In* H. Vagtborg (ed.), *The baboon in medical research*. Austin: Univ. Texas, 1965. Pp. 65-80.

Lahiri, R. K., & Southwick, C. H. Parental care in *Macaca sylvana*. *Folia Primatologica*, 1966, **4**, 257-264.

Lawick-Goodall, Jane van. Mother-offspring relationships in free-ranging chimpanzees. *In* D. Morris (ed.), *Primate ethology*. London: Weidenfeld & Nicolson, 1967. Pp. 287-346.

Lawick-Goodall, Jane van. The behaviour of free-living chimpanzees in the Gombe Stream Reserve. *Animal Behavior Monographs*, 1968, **1**, 161-311.

Loizos, C. Play behaviour in higher primates: A review. *In* D. Morris (ed.), *Primate ethology*. London: Weidenfeld & Nicolson, 1967. Pp. 176-218.

Marsden, H. M. Agonistic behaviour of young rhesus monkeys after changes induced in social rank of their mothers. *Animal Behaviour*, 1968, **16**, 38-44.

Martin, R. D. Reproduction and ontogeny in tree-shrews (*Tupaia belangeri*). *Zeitschrift für Tierpsychologie*, 1968, **25**, 409-495, 505-532.

Mason, W. A. Development of communication between young rhesus monkeys. *Science*, 1959, **130**, 712-713.

Mason, W. A. The effects of social restriction on the behavior of rhesus monkeys: I. Free social behavior. *Journal of Comparative and Physiological Psychology*, 1960, **53**, 582-589.

Mason, W. A. The effects of social restriction on the behavior of rhesus monkeys: II. Tests of gregariousness. *Journal of Comparative and Physiological Psychology*, 1961, **54**, 287-290. (a)

Mason, W. A. The effects of social restriction on the behavior of rhesus monkeys: III. Dominance tests. *Journal of Comparative and Physiological Psychology*, 1961, **54**, 694-699. (b)

Mason, W. A. The effects of social restriction on the behavior of rhesus monkeys: IV. Responses to a novel environment and to an alien species. *Journal of Comparative and Physiological Psychology*, 1962, **55**, 363-368.

Mason, W. A. Sociability and social organization in monkeys and apes. *In* L. Berkowitz (ed.), *Recent advances in experimental social pyschology*. New York: Academic Press, 1964. Pp. 277-305.

Mason, W. A. Determinants of social behavior in young chimpanzees. *In* A. M. Schrier, H. F. Harlow, & F. Stollnitz (eds), *Behavior of nonhuman primates*. Vol. II. New York: Academic Press, 1965. Pp. 335-364. (a)

Mason, W. A. The social development of monkeys and apes. *In* I. DeVore (ed.), *Primate behavior*. New York: Holt, Rinehart & Winston, 1965. Pp. 514-543. (b)

Mason, W. A. Motivational aspects of social responsiveness in young chimpanzees. *In* H. W. Stevenson, E. H. Hess, & Harriet L. Rheingold (eds), *Early behavior: Comparative and developmental approaches*. New York: Wiley, 1967. Pp. 103-126.

Mason, W. A., & Hollis, J. H. Communication between young rhesus monkeys. *Animal Behaviour*, 1962, **10**, 211-221.

Meier, G. W. Other data on the effects of social isolation during rearing upon adult reproductive behaviour in the rhesus monkey. *Animal Behaviour*, 1965, **13**, 228-231.

Missakian, Elizabeth A. Reproductive behavior of socially deprived rhesus monkeys. *Journal of Comparative and Physiological Psychology*, 1969, **69**, 403-407.

Mitchell, G. D. Intercorrelations of maternal and infant behaviors in *Macaca mulatta*. *Primates*, 1968, **9**, 85-92. (a)

Mitchell, G. D. Persistent behavior pathology in rhesus monkeys following early social isolation. *Folia Primatologica*, 1968, **8**, 132-147. (b)

Mitchell, G. D., Arling, G. L., & Møller, G. W. Long-term effects of maternal punishment on the behavior of monkeys. *Psychonomic Science*, 1967, **8**, 205-210.

Mitchell, G. D., Harlow, H. F., Griffin, G. A., & Møller, G. W. Repeated maternal separation in the monkey. *Psychonomic Science,* 1967, **8,** 197-198.

Mitchell, G. D., Raymond, E. J., Ruppenthal, G. C., & Harlow, H. F. Long-term effects of total social isolation upon behavior of rhesus monkeys. *Psychological Reports,* 1966, **18,** 567-580.

Møller, G. W., Harlow, H. F., & Mitchell, G. D. Factors affecting agonistic communication in rhesus monkeys (*Macaca mulatta*). *Behaviour,* 1968, **31,** 339-357.

Moog, G. Geburt und Aufzucht einer Gelbgrüner Meerkatze *Cercopithecus callitrichus. Zoologische Garten, Leipzig,* 1957, **23,** 220-223.

Moynihan, M. Some behavior patterns of platyrrhine monkeys: I. The night monkey (*Aotus trivirgatus*). *Smithsonian Miscellaneous Collections,* 1964, **146,** 5.

Napier, J. R., & Napier, P. H. *A handbook of living primates.* New York: Academic Press, 1967.

Nolte, A. & Dücker, G. Jugendentwicklung eines Kapuzineraffen (*Cebus appela,* L.) mit besonderer Berücksichtigung des Wechselseitigen Verhaltens von Mutter und Kind. *Behaviour,* 1959, **14,** 335-373.

Ploog, D. W. Biological bases for instinct and behavior. *In* J. Wortis (ed.), *Recent advances in biological psychiatry.* Vol. 8. New York: Plenum Press, 1966. Pp. 199-203.

Ploog, D. W. The behavior of squirrel monkeys (*Saimiri sciureus*) as revealed by sociometry, bioacoustics, and brain stimulation. *In* S. A. Altmann (ed.), *Social communication among primates.* Chicago: Univ. Chicago Press, 1967. Pp. 149-184.

Plutchnik, R. The study of social behaviour in primates. *Folia Primatologica,* 1964, **2,** 67-92.

Pournelle, G. H. Observations on the birth and early development of Allen's monkey. *Journal of Mammalogy,* 1962, **43,** 265-266.

Randolph, M. C., & Mason, W. A. Effects of rearing conditions on distress vocalizations in chimpanzees. *Folia Primatologica,* 1969, **10,** 103-112.

Raphael, D. Uncle rhesus, auntie pachyderm, and mom: All sorts and kinds of mothering. *Perspectives in Biology and Medicine,* 1969, **12,** 290-297.

Reynolds, V., & Reynolds, Frances. Chimpanzees of the Budongo Forest. *In* I. DeVore (ed.), *Primate behavior.* New York: Holt, Rinehart & Winston, 1965. Pp. 368-424.

Rioch, D. McK. Discussion of agonistic behavior. *In* S. A. Altmann (ed.), *Social communication among primates,* Chicago: Univ. Chicago Press, 1967. Pp. 115-122.

Rosenblum, L. A. *The development of social behavior in the rhesus monkey.* (Doctoral dissertation, University of Winconsin) Ann Arbor, Mich.: University Microfilms, 1961. No. 61-3158.

Rosenblum, L. A. Mother–infant relations and early behavioral development in squirrel monkey. *In* L. A. Rosenblum & R. W. Cooper (eds), *The squirrel monkey.* New York: Academic Press, 1968. Pp. 207-233.

Rosenblum, L. A., & Harlow, H. F. Approach-avoidance conflict in the mother-surrogate situation. *Psychological Reports,* 1963, **12,** 83-85.

Rosenblum, L. A., & Kaufman, I. C. Laboratory observations of early mother–infant relations in pigtail and bonnet macaques. *In* S. A. Altmann (ed.), *Social communication among primates.* Chicago: Univ. Chicago Press, 1967. Pp. 23-41.

Rosenblum, L.A., Kaufman, I. C., & Stynes, A. J. Individual distance in two species of macaque. *Animal Behaviour,* 1964, **12,** 338-342.

Rowell, T. E. Some observations on a hand-reared baboon. *In* B. M. Foss (ed.), *Determinants of infant behaviour.* Vol. III. London: Methuen, 1965. Pp. 77-82.

Rowell, T. E. Hierarchy in the organization of a captive baboon group. *Animal Behaviour,* 1966, **14,** 430-443.

Rowell, T. E. A quantitative comparison of the behaviour of a wild and a caged baboon group. *Animal Behaviour,* 1967, **15,** 499-509.

Rowell, T. E. The effect of temporary separation from their mothers on the mother–infant relationship of baboons. *Folia Primatologica,* 1968, **9,** 114-122.

Rowell, T. E., Din, N. A., & Omar, A. The social development of baboons in their first three months. *Journal of Zoology,* 1968, **155,** 461-484.

Rowell, T. E., Hinde, R. A., & Spencer-Booth, Y. "Aunt"–infant interaction in captive rhesus monkeys. *Animal Behaviour,* 1964, **12,** 219-226.

Rumbaugh, D. M. Maternal care in relation to infant behavior in the squirrel monkey. *Psychological Reports,* 1965, **16,** 171-176.

Sackett, G. P. The persistence of abnormal behavior in monkeys following isolation rearing. *In* Ruth Porter (ed.), *The role of learning in psychotherapy.* London: Churchill, 1968. Pp. 3-25. (a)

Sackett, G. P. Abnormal behavior in laboratory-reared rhesus monkeys. *In* M. W. Fox (ed.), *Abnormal behavior in animals.* Philadelphia: Saunders, 1968. Pp. 293-331. (b)

Sade, D. S. Some aspects of parent-offspring and sibling relations in a group of rhesus monkeys, with a discussion of grooming. *American Journal of Physical Anthropology,* 1965, **23,** 1-18.

Sade, D. S. Determinants of dominance in a group of free-ranging monkeys. *In* S. A. Altmann (ed.), *Social communication among primates.* Chicago: Univ. Chicago Press, 1967. Pp. 99-114.

Sauer, E. G. F. Mother–infant relationship in galagos and the oral child-transport among primates. *Folia Primatologica,* 1967, **7,** 127-149.

Schaller, G. B. *The Mountain Gorilla.* Chicago: Univ. Chicago Press, 1963.

Schaller, G. B. The behavior of the mountain gorilla. *In* I. DeVore (ed.), *Primate behavior.* New York: Holt, Rinehart & Winston, 1965. Pp. 324-367.

Schlott, M. Notizen zur Kenntnis der Jugendentwicklung der Gelbrünen Meerkatze (*Cercopithecus callitrichus* ls Geoffroy). *Zoologische Garten, Leipzig,* 1956, **21,** 270-274.

Seay, B. Maternal behavior in primiparous and multiparous rhesus monkeys. *Folia Primatologica,* 1966, **4,** 146-168.

Seay, B., Alexander, B. K., & Harlow, H. F. Maternal behavior of socially deprived rhesus monkeys. *Journal of Abnormal and Social Psychology,* 1964, **69,** 345-354.

Seay, B., Hansen, E., & Harlow, H. F. Mother–infant separation in monkeys. *Journal of Child Psychology and Psychiatry and Allied Disciplines,* 1962, **3,** 123-132.

Seay, B., & Harlow, H. F. Maternal separation in the rhesus monkey. *Journal of Nervous and Mental Diseases,* 1965, **140,** 434-441.

Simonds, P. E. The bonnet macaque in South India. *In* I. DeVore (ed.), *Primate behavior.* New York: Holt, Rinehart & Winston, 1965. Pp. 175-196.

Southwick, C. H., Beg, M. A., & Siddiqi, M. R. Rhesus monkeys in North India. *In* I. DeVore (ed.) *Primate behavior.* New York: Holt, Rinehart & Winston, 1965. Pp. 111-159.

Spencer-Booth, Y. The behaviour of group companions towards rhesus monkey infants. *Animal Behaviour,* 1968, **16,** 541-557. (a)

Spencer-Booth, Y. The behaviour of twin rhesus monkeys and comparisons with the behaviour of single infants. *Primates,* 1968, **9,** 75-84. (b)

Spencer-Booth, Y. The effects of rearing rhesus monkey infants in isolation with their mothers on their subsequent behaviour in a group situation. *Mammalia,* 1969, **33,** 80-86.

Spencer-Booth, Y., & Hinde, R. A. The effects of separating rhesus monkey infants from their mothers for six days. *Journal of Child Psychology and Psychiatry and Allied Disciplines,* 1967, **7,** 179-197.

Spencer-Booth, Y., & Hinde, R. A. Tests of behavioural characteristics for rhesus monkeys. *Behaviour,* 1969, **33,** 179-211.

Spencer-Booth, Y., & Hinde, R. A. Effects of 6-days separation from mother on 18- to 32-week old rhesus monkeys. *Animal Behaviour,* 1971, **19,** 174-191.

Struhsaker, T. T. Social structure among vervet monkeys (*Cercopithecus aethiops*). *Behaviour,* 1967, **29,** 83-121.

Struhsaker, T. T. Correlates of ecology and social organization among African cercopithecines. *Folia Primatologica,* 1969, **11,** 80-118.

Stynes, A. J., Rosenblum, L. A., & Kaufman, I. C. The dominant male and behavior within heterospecific monkey groups. *Folia Primatologica,* 1968, **9,** 123-134.

Sugiyama, Y. Behavioural development and social structure in two troops of hanuman langurs. (*Presbytis entellus*). *Primates,* 1965, **6,** 213-247. (a)

Sugiyama, Y. On the social change of hanuman langurs (*Presbytis entellus*) in their natural condition. *Primates,* 1965, **6,** 381-418. (b)

Sugiyama, Y. Social organization of hanuman langurs. *In* S. A. Altmann (ed.), *Social communication among primates.* Chicago: Univ. Chicago Press, 1967. Pp. 221-236.

Tanaka, J. Social structure of Nilgiri langurs. *Primates,* 1965, **6,** 107-122.

Tsumori, A. Newly acquired behavior and social interactions of Japanese monkeys. *In* S. A. Altmann (ed.) *Social communication among primates.* Chicago: Univ. Chicago Press, 1967. Pp. 207-219.

Vandenbergh, J. G. Behavioral observations of an infant squirrel monkey. *Psychological Reports,* 1966, **18,** 683-688.

Vandenbergh, J. G. The development of social structure in free-ranging rhesus monkeys. *Behaviour,* 1967, **29,** 179-194.

Weiskrantz, L., & Cowey, A. The aetiology of food reward in monkeys. *Animal Behaviour,* 1963, **11,** 225-234.

Wilson, A. P., & Vessey, S. H. Behavior of free-ranging castrated rhesus monkeys. *Folia Primatologica,* 1968, **9,** 1-14.

Yamada, M. A study of blood-relationship in the natural society of the Japanese macaque. *Primates,* 1963, **4,** 43-65.

Yoshiba, K. Local and intertroop variability in ecology and social behavior of common Indian langurs. *In* Phyllis C. Jay (ed.), *Primates.* New York: Holt, Rinehart & Winston, 1968. Pp. 217-242.

Chapter 2

Activity Profiles of Primate Groups[1]

Irwin S. Bernstein

Yerkes Regional Primate Research Center, Emory University,
Atlanta, Georgia

I. INTRODUCTION

When the subjects live as members of organized groups, another dimension of complexity is introduced to the study of animal behavior. The existence of a society, as opposed to an aggregation, implies the presence of social mechanisms that operate to maintain a particular social order. Investigation of these social mechanisms requires additional abstraction from descriptions of individual activities and the responses of one individual to another. The organization of the social unit may pervasively influence almost all individual and interindividual

[1] This research was supported by U.S. Public Health Service grants MH-13864 from the National Institute of Mental Health and FR-00165 from the National Institutes of Health, Animal Resources Branch.

activities. The existence of a society implies influences upon each member from the group as a whole.

To begin the study of a group, one must identify the basic elements of behavior—the response units. Frequencies and durations of each response element can then be measured. Analysis of the data for response sequences will yield information on consistent patterns, and suggest the function of the responses. With a knowledge of response patterns in the group, and with possible response interpretations, an investigator can postulate roles, role patterns, and social mechanisms that unify the animals into an organized unit or society. Such a model can then be used to understand the interrelationships among members of a society and the influences that the society has on each of its members. Likely outcomes in specified situations can be predicted by examining consistent role patterns and the social mechanisms that will be elicited in the situations.

Although final evaluation of the model depends on completing the four steps outlined (identifying response elements, measuring frequencies and durations of response elements, analyzing response sequences, postulating social mechanisms and roles in the society), each of the four steps will provide information concerning the society under study.

The definition of a society may be intuitively obvious, yet, it is difficult to verbalize. Sarbin (1954) defines a society as a set of interrelated roles. Altmann (1962) defines the social unit by its communication network. Differential frequencies of communication exchanges demarcate the boundaries of the unit. The differential frequencies of the responses themselves, however, may reflect the nature of the social organization.

The nature of the social organization may be presumed to have been shaped by the ecological pressures of the natural habitat in which the individuals and their social organization evolved. For example, the particular forms of communication responses and their frequencies will reflect the communication requirements necessary to establish, maintain, and coordinate a social unit living under the specific ecological conditions that prevail in the habitat where a species evolved. Similarly, each of the response elements in the total behavioral repertoire is influenced by selective pressures in the natural habitat and the requirements of the particular social milieu. The form and frequency of response elements will thus reflect the selective pressures operating in the natural habitat of a species and the organizational features characteristic of the social units of the taxon. The behavior and morphology of an animal are so interrelated that both should reflect the special adaptations of a species to its environment. Both the form of response expressions and the total activity profile, as obtained by measuring the frequencies and durations of identifiable activities, will, therefore, reflect species-specific adaptations. This is not to say that each element will be distinctive, any more than each morphological feature need be distinctive, for the processes of convergent or parallel evolution, or both, may apply to behavior as well as morphology. Instead, it is claimed that the activity profile represents the total behavioral adaptation and that the behavioral and morphological adaptations of a species define the characteristics of that species.

Almost all of the monkeys and apes studied so far live as members of a social unit. In fact, this seems to be one of the few universal generalizations about primate behavior that has persisted as our knowledge of different monkey and ape taxa has expanded. To be sure, some individuals may live alone for relatively brief periods, but in general, monkeys and apes are found as members of a stable social unit. The social unit may vary considerably from taxon to taxon, and, even within a species, ecological pressures may exert strong influences on the expression of social organization.

Social units reported include: a single pair and their offspring,[2] a single male with multiple adult females and young,[3] multiple males with multiple females and young,[4] one-male units (harems) which associate together to form larger bands,[5] and fluid reassociation of individuals within a fixed community.[6] In addition, social units may be intolerant of other conspecific social units, as in territorial species [e.g., titi monkeys, Mason, 1968; lutong *(Presbytis cristatus)*, Bernstein, 1968], or they may express some degree of tolerance for other social units, e.g., troops of gorillas *(Gorilla gorilla)* may meet, mingle, pass the night together, and then move their separate ways (Schaller, 1963, 1965). Hostility towards other social units, conspecific individuals, or both, may be related to territoriality, but may be influenced by a variety of other social mechanisms.

This chapter examines the behavior of primates living in social groups. Comparisons among several selected taxa will be based on differential activity profiles obtained through the measurement of frequencies and durations of identified activities. The work of Davis, Leary, Casebeer Smith, and Thompson (1968) has demonstrated that activity-pattern profiles may be used to identify species membership of individual animals in that the profiles obtained from only a limited number of activities were distinctive for each of the seven taxa studied. Of course, temporary stresses and environmental variables may influence these response rates, and it has already been demonstrated that groups within a species

[2] For example, gibbons *(Hylobates lar)* (Carpenter, 1940; Ellefson, 1968); titi monkeys *(Callicebus moloch)* (Mason, 1966).

[3] For example, some langurs *(Presbytis* spp.) (Bernstein, 1968; Poirier, 1969; Sugiyama, 1967); patas monkeys *(Erythrocebus patas)* (Hall, 1965).

[4] For example, most macaques *(Macaca* spp.) (Bernstein, 1967a; Simonds, 1965; Southwick, Beg, & Siddiqi, 1965); savannah baboons *(Papio* spp.) (Hall & DeVore, 1965); howler monkeys *(Alouatta villosa)* (Bernstein, 1964a; Carpenter, 1934; Collias & Southwick, 1952).

[5] For example, gelada baboons *(Theropithecus gelada)* (Crook, 1966); hamadryas baboons *(Papio hamadryas)* (Kummer, 1968).

[6] Chimpanzees *(Pan troglodytes)* (van Lawick-Goodall, 1968; Reynolds, 1963; Reynolds & Reynolds, 1965).

(Gartlan & Brain, 1968; Hall, 1963), and in fact even the same animals at different times (Crook, 1967; Kummer, 1968), can modify their basic social organization in response to stress and environmental demands. Minor fluctuations in activity profiles may, therefore, reflect social modifications in response to environmental pressures. Constant small fluctuations in a stable environment should reflect the plasticity that allows ready modification to meet changing conditions. Nevertheless, species-typical patterns should persist and different primate taxa should respond differentially to the same environmental demands. The total activity profile will reflect the behavioral adaptations of a particular taxon, but any particular response may occur in multiple taxa. Once again, the behavioral and morphological adaptations of a species are so interdependent that they must be considered as one and inseparable. To the extent that one may be expected to define a species, so should the other.

Certain elements of primate social organization appear to be stable and conservative even in the face of minor fluctuations in environmental variables. If we define a society as a collection of interrelated roles (Bernstein & Sharpe, 1965; Sarbin, 1954), then the social organization may be viewed as stable so long as the various roles, or consistent patterns of responses, can be identified within a unit. Since more than one role can be filled by an individual, and since more than one individual may fulfill a particular role function, measurements of "roles" are difficult. They may be defined and recognized, but it would be difficult to define social organizations by differential role patterns, except insofar as a presumed crucial role could be demonstrated to be present or absent. An indirect objective measure of a role in primate societies is the measurement of frequencies of associated patterns of activities. Such an approach appears promising, but the measurement and analysis of sequential events is exceedingly laborious. Such sequential information is, however, the next logical step after compiling information on (1) the form of response elements and (2) the frequency rates and duration of such responses.

Other approaches to the study of primate societies have concentrated on a single aspect such as dominance and dominance hierarchies. Some have hypothesized that this single characteristic is present in all primate societies, but whereas its presence is convincingly demonstrated in most macaques and baboons, conclusive evidence is lacking for New World monkeys (Ceboidea), leaf monkeys (Colobinae), and apes (Hominoidea). In these taxa, proponents of universal status-hierarchy models insist that the gradient is of small slope, whereas other investigators find no evidence for status relationships at least among some categories of group members. Measurement of the slope of dominance gradients and/or the degree of linearity might prove an important comparative approach, but such measurement must await the development of more sophisticated testing techniques. At present, we cannot even assume status gradients have pervasive influences on other aspects of social organization

(Bernstein, 1970b). Models of status relationships based on measures of priority to presumed incentives may be questioned, and the use of this one measure to investigate primate societies seems to oversimplify the interrelationships of individuals in such a society. (See Bernstein, 1970b, for a more complete review and discussion; also see Chapter 1 by Hinde.)

A. Field and Laboratory Approaches

In addition to multiple theoretical models of social organization, investigators have also been faced with two basic study techniques, the field study and the laboratory study. Each has strong points and weaknesses, but fortunately the two approaches are supplementary and complementary. The field study ideally focuses on undisturbed animals living in the intact natural habitat in which they evolved. Although such idyllic conditions seldom exist, certain general ecological features must prevail for a habitat to be called "natural." Minimum standards have been proposed (Bernstein, 1967b). The natural habitat should present the whole Gestalt permitting study of such factors as: migration, intertroop responses, territoriality, seasonality, home range, day range, food use, predator defense, as well as the social organization and activity patterns of the troop. Many of these factors can only be studied under field conditions. Field conditions, however, are also beset with inherent limitations. The subjects control the situation, and the investigator can only minimize his disruption of the natural environment. Under forest conditions, visibility may become a major problem; replication of data collected under unusual conditions may be impossible; and there can be little control of significant variables. It is exactly these areas, replication and control, that are the strengths of laboratory studies. The two techniques must, therefore, be used to supplement each other.

Laboratory study situations have been criticized as barren and artificial environments that distort behavior. Behavior is, however, always in response to existing conditions, and subjects can only respond with patterns already available to them. Thus, the laboratory may readily be used to test the limits of response adaptation to extreme environments where selected variables can be controlled.

Bourlière (1961) has pointed to the common erroneous assumption that the value of data is directly related to the difficulty of the conditions under which it was obtained. Other field investigators have correctly recognized the variations in response patterns and social organization found in the same species at different study sites and fear that if activity patterns are sensitive to such small ecological differences, then the conditions that prevail in a laboratory would drastically distort behavior. To allay such fears, one need only examine the findings of experimenters who have compared data collected on the same species in both field and laboratory studies. Bernstein (1967a), Eisenberg and Kuehn

(1966), Gartlan and Brain (1968), Kummer and Kurt (1965), Rowell (1967), and Southwick (1967), among others, have all reported that the expression of responses in the species studied were virtually identical under field and captive conditions, although the frequencies of such expressions have varied considerably. In fact, captive conditions enhance the opportunity to study social behavior, because the frequency of social activities rises when primates gain leisure time through the elimination of the necessity to spend long hours foraging for food. Even in the wild, studies have shown that social activities decrease under impoverished conditions (Gartlan & Brain, 1968). Captive primates, with many activities limited and all of the necessities provided, must spend more time occupied in the activities that remain available. Social interactions inevitably increase in a confined space, and it may well be these responses, and their range of expression, that an investigator wishes to study.

Many investigators have tried to achieve the best of both study techniques through the use of compromise situations. Groups have been transplanted to habitats where conditions are more favorable for study; island colonies have been established, and enclosures, compounds, and large cages have been stocked with selected animals. Such solutions permit considerable control over the subjects, ready replication, and easy visibility. The limitations of such conditions can be readily appreciated, but long-term studies of appropriate social processes can be undertaken.

B. An Approach to the Study of Captive Groups

My approach to the study of primate social organizations has made extensive use of captive groups living in outdoor compounds. Specific experimental manipulations are used to examine selected problems, and results are cross checked against data obtained, using similar data-collecting techniques, during field studies of the same and other taxa.

The first problem in collecting quantitative data on the activities of a captive group is to identify both the animals and the responses that one wishes to study. Meaningful response units must be selected and reliable means of scoring such responses must be devised. The net result of studies of the taxonomy of behavior is a kind of dictionary describing motor patterns, postures, expressions, and vocalizations that are characteristic of various subclasses of animals. The contents of this dictionary are thus determined by the following factors: the expression of responses in individuals under observation, their ready identification by the observer, and the theoretical orientation of the observer. This last factor may result in a decision to include one facial expression because of its presumed communicative significance, and not include a second facial expression because that expression is viewed as an individual response having no special significance. Although it might seem that the vocabularies developed by different investigators would vary widely, and that the identified response units might be arbitrarily selected for inclusion or exclusion, in fact, most investigators have identified much the same sets of responses, and at the same level of analysis, despite differences in the labels used for the definitions. The responses to be discussed in this chapter have been selected from a larger vocabulary used to compare many primate taxa.

Species-specific responses and responses common only to a limited number of related species have special taxonomic interest. The taxonomic significance of precise response expressions is discussed by Bernstein (1970c). Responses found in a broad range of taxa, and the more grossly-defined response categories, were used to test for quantitative systematic changes in the frequency and duration of such responses in selected taxa. The balance of this chapter will be used to describe one possible analysis demonstrating quantitative differences in response profiles.

New data were collected on seven groups representing six primate taxa. Similar data from other groups representing one of these plus three other taxa were included in comparisons. Data were collected by a time-sampling technique wherein each animal in each group was observed at specified times for a fixed amount of time. A balanced order was used to ensure that equivalent data were collected on all animals in all groups. Because of diurnal cyclical changes, seasonal and weather effects, data were collected at selected hours of the day under prescribed weather conditions from animals living in nearly identical compounds. Data collection provided measures of the frequency rates and the percentage of time individuals were engaged in predefined activities. Most of the data concern visually observable responses—analysis of auditory components was not attempted.

It was hypothesized not only that the activity profiles would be similar when taxonomically closely-related species were compared, but that the degree of phylogenetic affinity would be reflected in the degree of similarity in activity profiles. Possible parallels might also be found for species adapted to similar habitats, but the activity profiles should be capable of differentiating these forms as well as congeneric forms adapted to different habitats. Furthermore, not only should the profiles differentiate, but they should also indicate more similarity between congeneric forms than between more distantly related taxa. The extent to which age, sex, and similar variables also affect activity profiles is not known; should these effects prove to mask taxonomic affiliations, future research will require control of these variables.

II. METHOD

Seven groups, representing six primate taxa, were maintained with minimal disturbance in outdoor compounds at the Yerkes Field Station in northern

Georgia. Four groups were housed in compounds with 100- x 100-ft outdoor areas and 30 x 10-ft attached indoor quarters always available to the animals. The other three groups had 50- x 50-ft outdoor areas with 10- x 15-ft attached indoor quarters. The outdoor areas were unroofed and initially included the natural vegetation of northern Georgia. Many of the trees and other plants were, however, destroyed by the animals within a few weeks of release into these areas.

Each compound had an attached observation post from which every section of both indoor and outdoor quarters could be seen without the observers having to move.

A. Groups

The groups were formed initially by attempting to approximate the natural sex composition of small troops. Wherever possible, the age composition of natural troops was also approximated. All groups were maintained with minimal disturbance after formation, and all six taxa had produced numerous infants within a year of group formation. No attempt was made to manipulate age or sex compositions of the groups after formation, and final composition was the result of natality and mortality. Data on group size, dates of group formation, and inclusive dates of data collection are shown in Table I.

The gelada group *(Theropithecus gelada)* comprising three adult males and eight adult females was formed in May 1968. Two harems were established with the bachelor male attaching himself to the smaller harem. Five births occurred during the study period, and several females were pregnant at the completion of this study.

The pigtailed monkey group *(Macaca nemestrina)* was formed in 1963, but animals were added and removed until 1964. The oldest animals born into the group produced infants during the spring of 1969. Data were collected on this group intermittently from the time of group formation (Bernstein, 1966, 1969a, b, 1970a); information collected under previous conditions is reported separately.

The group of West African green monkeys *(Cercopithecus sabaeus)* consisted of nine animals in August 1968 when data collection started. The oldest male and female were the parents of all of the other animals, and the group had been maintained intact since the birth of the first infant. During the course of the study year, one infant was born, but died due to infection after compound fracture of the left leg. The original breeding pair also died, primarily as a result of periodic severe fighting with their newly adult offspring.

The cynomolgus monkeys *(Macaca fascicularis)* were a group obtained from the Monkey Jungle south of Miami. The owner, Frank DuMond, stated that six animals were purchased in 1933, and that his present colony of 150 semifree-ranging animals are all descendants of these animals. The group in the

TABLE I

Subject Groups and Data-Collection Dates

Data	Pigtailed macaque	Cynomolgus macaque	Celebes	Gelada	Mangabey 1	Mangabey 2	Green monkey
Date of group formation	September 1962	September 1968	August 1968	May 1968	September 1968	February 1969	September 1962
Data collection started	July 1968	March 1969	September 1968	May 1968	November 1968	May 1969	August 1968
Data collection completed	December 1968	July 1969	May 1969	February 1969	May 1969	July 1969	July 1969
Group size, collection started	25	26	12	11	9	7	9
Group size, at completion	27	27	13	13	9	7	7

present study had resulted from spontaneous fissioning of the main group. These animals were trapped in August 1968 and transplanted to the Yerkes Field Station. Trapping operations netted 24 animals with only perhaps one or two individuals escaping the process. The group rapidly adapted to their new quarters and six infants were born during the first year.

The group of Celebes black apes *(Cynopithecus niger)* included a juvenile male Celebes or moor macaque *(Macaca maurus)* for reasons unrelated to this study. He appeared to have become a well-integrated member of the group, and it would be difficult to say to what extent his presence influenced the Celebes black apes. The two groups of sooty mangabeys *(Cercocebus atys)* initially consisted of fully adult animals. These two groups and the Celebes group were formed just prior to data collection, but were judged to have already stabilized social relations. By the first year after group formation, three infants had been born into the Celebes group, and four infants into the first mangabey group. The second mangabey group had been in existence for only 6 months, but several females were pregnant by the end of the study.

All animals were selected as intact healthy specimens at the time of group formation, with the following exceptions: (1) The canine teeth of two male geladas had been rounded and blunted several years previously. These teeth were still quite large and can best be described as artificially aged. (2) The canines of the five male Celebes black apes had been similarly modified.

B. Data-Collection Procedures

Data collection consisted of four ten-session blocks of observations conducted on each of the study groups, except the cynomolgus macaque group and the second mangabey group for each of which only two blocks were completed. During each of the ten sessions in a block, each animal was observed by two observers for ten 30-sec periods. Two checklists of selected activities and social interactions were used. One observer marked each time selected responses were seen, while the second observer scored the occurrence or nonoccurrence of selected responses during each 30-sec period. Animals interacting with the animal being observed (subject), and the directionality of responses, were scored whenever appropriate. Only those responses involving the subject either as actor or recipient were scored during data-collection periods for that subject. In addition, a summary of group behavior was obtained each session by counting the total number of group members involved in selected broad categories of behavior during ten 30-sec periods. Individual and group data were analyzed to show typical activity patterns during the block of ten data-collection sessions, as well as individual totals and subtotals by sex. Scores differing by 10% or more probably indicate reliable differences, since data collected on a single group using these procedures differed less than 10% between two 10-day blocks (Bernstein & Draper, 1964).

Two data blocks were composed of sessions starting late in the morning and two data blocks were composed of sessions starting early in the afternoon. No more than one session was scheduled per day. Data were collected only on days when the air temperature ranged between 50 and 80°F, and when the sky was clear or partly cloudy. Observations were conducted from the observation window after all animals were thoroughly habituated to the persons scoring the group. Habituation was achieved by having the observers work all day in the observation room doing a variety of other quiet tasks and making informal observations and photographic records.

Data for rhesus monkeys *(Macaca mulatta)*, cebus monkeys *(Cebus albifrons)*, and gibbons *(Hylobates lar)* in concrete-floored vegetationless pens had been obtained in previous studies (Bernstein, 1964b, 1965; Bernstein & Mason, 1963; Bernstein & Schusterman, 1964). Data collected on another Celebes group and on the same pigtailed macaque group under the conditions prevailing for the rhesus, cebus, and gibbon groups were also available. When compared to data collected in the present study, these data suggested situational effects for many response categories (see Section III, B).

One broad influence operating in the case of the rhesus monkeys (two distinct groups) was high temperature. The data for these groups were obtained when temperatures were between 65 and 95°F as compared with the 80°F maximum adhered to in the present study. Temperature effects were analyzed and reported (Bernstein & Mason, 1963). These effects would indicate that the rhesus scores differ from those presented for the other macaques in precisely those activities previously demonstrated to be influenced by high temperatures.

III. RESULTS AND DISCUSSION

A. Biases in Scoring

The difficulties inherent in scoring the activities of an entire group simultaneously can be detected by comparing Tables II and III. Scoring whole groups yielded the data in Table II, which shows the percentage of group members occupied in the activities listed during 30-sec periods. In contrast, individual scoring sessions yielded the percentage of 30-sec periods in which individuals in each group were occupied in the activities listed in Table III.

Despite excellent visibility from the observation point, an investigator scanning a group spread over the area of one of the compounds often failed to notice brief responses involving only one or two animals, unless these responses were accompanied by loud vocalizations or dramatic movements. These observational failures increase both with group size and with the area occupied.

TABLE II

Mean Percentage of Group Members Engaged in Selected Activities During 30-Second Periods and Locations at Onset of Periods

Activity or Location	Rhesus	Pigtails	*Pigtails*	*Cynomolgus*	Celebes	Geladas	Mangabeys 1	*Mangabeys 2*	*Greens*	Cebus	Gibbons
Social grooming	15	16	9	17	18	26	9	21	7	6	2
Play	+	4	4	10	5	1	2	–	8	+	22
Agonistic responses	2	2	2	3	5	3	2	3	2	1	–
Sexual responses	1	3	1	1	2	4	4	3	+	–	–
Inactive	26	8	12	2	11	14	9	9	12	5	12
More than 6 feet above ground	NA	NA	4	2	3	3	1	–	14	NA	NA
On the ground	NA	NA	82	93	76	86	70	74	63	NA	NA
Travel	NA	34	42	34	34	17	30	20	30	30	NA
Drinking.	2	2	3	1	2	1	2	2	1	1	6
Eating	8	27	19	17	25	23	36	26	17	18	26
Object manipulation	1	2	2	1	3	4	6	21	2	1	NA
Self-directed activity	9	14	5	10	13	16	15	26	10	10	NA

NOTE: NA means the score was not available or not applicable. A + indicates score was less than 0.5%, whereas – indicates no occurrences were scored. *Italic* group names indicate groups included in the present study.

TABLE III

Mean Percentage of 30-Second Periods in Which Individual Was Seen in Selected Activities and Locations

Activity or Location	Rhesus	Pigtails	*Pigtails*	*Cynomolgus*	*Celebes*	*Geladas*	*Mangabeys 1*	*Mangabeys 2*	*Greens*	*Cebus*	*Gibbons*
Proximity to others	57	59	63	70	68	74	51	43	57	66	66
Nonspecific contact	14	6	16	20	12	20	7	1	23	29	38
Social grooming	21	24	17	26	22	37	15	23	9	6	4
Grooming as percentage of proximity	37	43	27	38	33	50	29	53	15	8	6
Play	1	3	6	9	5	1	1	–	7	+	28
Inactive	43	16	15	3	12	15	14	12	19	11	7
Indoors	NA	NA	8	2	21	–	34	29	9	NA	NA
Highest locations	NA	NA	3	3	4	5	1	–	10	NA	NA
On the ground	NA	NA	86	94	78	90	75	71	76	NA	NA
Travel	15	35	42	45	35	19	31	27	33	36	73
Drinking	1	3	3	2	3	2	3	4	2	1	4
Eating	6	25	26	36	19	16	31	23	24	23	33
Object manipulation	1	4	12	7	8	5	10	3	2	3	–
Self-directed activity	12	12	19	29	24	28	19	25	19	9	11

NOTE: NA means the score was not available or not applicable. A + indicates score was less than 0.5% whereas – indicates no occurrences were scored. *Italic* group names indicate groups included in the present study.

This bias has been discussed by Bernstein (1970a). It should be noted that the errors produced by group scoring are in the same categories as the discrepancies noted in estimating response frequencies using captive groups and wild troops under conditions of less favorable visibility. Despite a prior knowledge of the bias, and division of the group scoring chores among several observers, group data still tend to underestimate scores obtained by pooling individual scores in such categories as inactivity, drinking, manipulation of objects, self-directed activity, social grooming, and even total agonistic responses.

Inasmuch as situational effects have been shown to influence the exact frequencies of the activities shown in the tables, the specific measures shown may be expected to vary from measures obtained under other conditions. Care must be taken, then, in comparing data from different groups not collected under the conditions specified in this study. Vertical location scores, for example, were undoubtedly strongly influenced by the relative lack of trees with horizontal branches suitable for above-the-ground locomotion. This particular influence in the pigtailed monkey group has been described by Bernstein (1970a). The measures shown in Table II nonetheless indicate differences among taxa in utilization of the available elevated locations.

Two other possible biases in the data must be noted. Whereas biases attributed to the time of day that data were collected were controlled by using equal numbers of sessions run at two selected times, and whereas weather effects were controlled for, the age and sex composition of the groups was not controlled. Although all groups had multiple males and females (including adults of both sexes), the sex ratios varied unsystematically. That sex is a powerful influence on the nature and frequency of responses is amply demonstrated by the analysis of male and female activity profiles (see Section III, D). To control sex composition rigidly, however, would be to ignore the fact that differential sex ratios may be distinguishing attributes of different taxa. The variable of age composition, however, was also uncontrolled, and the short periods between group formation and data collection for most groups did not allow any natural adjustment of the age composition. Until further work stipulates age influences in these groups, we must consider this as a potential biasing factor.

B. Pigtailed Macaque Situation Effects

It should be noted that in the present study the pigtailed macaques were unremarkable with regard to contact scores, whereas in a previous study very low scores were noted, in agreement with the data of Rosenblum, Kaufman, and Stynes (1964). It may also be noted that the pigtailed macaque group spent relatively less time in sexual activities than had been true during previous studies. Movement to the new compound may account for these changes. Data under the two conditions have been compared and discussed by Bernstein (1970a) and are summarized in Tables II, III, and IV. The new area had natural vegetation and this apparently stimulated more object manipulation. Increases in self-directed grooming partially compensated for the apparent decrease in social grooming. Small changes in the other items in the activity profile may also be related to the living area, and additional testing is in progress. For the purposes of this study, however, it is reassuring to note that the two profiles for the pigtailed macaques are still remarkably similar to one another as compared to the other taxonomic groups represented.

C. Cross-Taxa Comparisons

1. SOCIAL RESPONSE CLASSIFICATIONS

One of the most basic measures of sociability is produced by recording spatial relations among animals. In the present case, two kinds of physical relations were noted: (1) the location of animals within 1 meter of the subject, and (2) nonspecific contact between any group member and the subject. Nonspecific contacts were differentiated from contacts that were part of any specifically identified social interaction, such as grooming, sexual behavior, play, or maternal care.

   a. Proximity. Simple proximity data might be expected to reflect maternal relations, thereby producing higher scores in those groups with extensive birth histories and large numbers of infants and young juveniles. Alternatively, in a restricted space, large groups might simply be crowded together. Neither explanation accounts for the highest scores being obtained in the gelada group (see Table III). Whereas members of this group spent three-fourths of their time within 1 meter of at least one other animal, in the other groups, animals typically spent two-thirds or less of their time in proximity to one another. The mangabeys tended to have the lowest scores, spending only about half the time in proximity to others, yet the social organization of the mangabey group was not remarkably different from the others. The scores in the gelada group were thus not a simple reflection of their harem type of social organization as opposed to the multimale organization characteristic of the other groups, but may nonetheless reflect one of the social mechanisms associated with one-male units.

   b. Contact. Nonspecific contact scores could not be simply attributed to age or maternal relationships inasmuch as the highest scores had been obtained in the gibbon and cebus groups. In these groups, despite the small number of maternal relationships, individuals spent nearly 40 and 30% of the time, respectively, in nonspecific contact with others (see Table III). The mangabeys had the lowest proximity scores and they also had the lowest nonspecific contact scores, tending to spend only about 5% of the time in nonspecific

TABLE IV

Mean Hourly Frequency of Individual Participation in Selected Activities

Activity	Rhesus	Pigtails	*Pigtails*	*Cynomolgus*	Celebes	Geladas	Mangabeys 1	*Mangabeys 2*	*Greens*	Cebus	Gibbons
Noncontact aggression	1	2	2	2	1	3	3	1	1	2	—
Contact aggression	+	2	2	1	1	+	1	1	1	1	—
Aggression (total)[a]	1	3	3	3	2	4	3	2	2	2	—
Noncontact (%)	92	53	47	77	56	90	76	67	41	61	NA
Submission	5	5	2	9	4	2	4	1	3	1	—
Agonistic (total)[a]	7	8	6	12	6	6	8	4	5	4	—
Submission (%)	79	59	41	74	69	32	56	42	57	31	NA
Mounting	1	6	2	1	1	1	1	1	+	+	—
Sexual soliciting	+	2	1	+	1	3	6	4	—	–	—
Genital investigation	NA	NA	1	3	1	2	2	1	+	NA	NA
Social sniffing	NA	NA	2	2	1	+	2	2	1	NA	NA
Lip-smacking	+	NA	1	2	6	4	1	–	2	+	NA
Species-specific patterns	NA	NA	4	1	2	1	4	3	1	NA	NA
All scored social responses	NA	NA	19	21	23	21	23	16	8	NA	NA

[a] Because of rounding, Aggression and Agonistic totals appear less than their components.

NOTE: NA means the score was not available or not applicable. A + indicates rate was less than 0.5, whereas – indicates no occurrences were scored. *Italic* group names indicate groups included in the present study.

contact with others. Contact scores for geladas, which had the highest proximity scores, however, were within the range of contact scores (12 to 23%) of the remaining groups in the present study.

   c. *Social grooming.* The social response that accounts for the greatest percentage of time that individuals are in contact with one another is social grooming. Maternal carriage and care, to be sure, account for higher scores with reference to individuals directly involved, but in all of the groups, grooming had the highest average score (Table III), despite relatively high natality figures in some groups.

   The geladas, with the highest proximity scores, also had the highest social grooming scores; grooming was recorded in 37% of the data periods. About half of the time geladas were in proximity with others, they were grooming them. Grooming scores in the gibbon, cebus, and green monkey groups were all less than 10%, and relatively little of their time in proximity with others was spent in grooming (6 to 15%). Individuals in the other groups all spent about 20 to 25% of their time in social grooming, although situational effects in the pigtailed monkeys (discussed in Section III, B) produced lower grooming scores (17%) during this study as compared with scores obtained in previous studies (24%), and animals in one of the mangabey groups spent only 15% of the time grooming. With the exception of lower percentages in the cebus, gibbon, and green monkey groups, as seen in Table III, 27 to 53% of proximities were accompanied by grooming.

   d. *Play.* Loizos (1967) has suggested that grooming serves to maintain sociality in groups characterized by strict dominance relationships, whereas play, especially among adults, may serve this function in other groups. The highest play scores (28%) did occur in the group with the lowest grooming scores, the gibbons, but the cebus group, with second-lowest grooming scores, showed one of the lowest scores for play (Table III). The cynomolgus macaques, with a score of 9%, claim second place for play, but this group also claims second place for grooming. The green monkey, pigtailed macaque, and Celebes groups all showed play scores in excess of 5%, and these groups all had a sizable number of juveniles present. An age effect is suggested, but the cebus group did have a number of young animals present. Further analyses for age influences will be required for more definitive answers. It should be noted, however, that play was observed even in the case of the highest-ranking adult males in the pigtailed macaque, cynomolgus macaque, and Celebes groups. The frequency of such behavior was low and was limited to periods when the groups seemed completely relaxed. However, its occurrence suggests that strong agonistic social orders do not preclude play by adult males of any social rank.

   e. *Agonistic responses.* Agonistic responses were scored and analyzed in some detail and then pooled to provide information on responses interpreted as (1) noncontact aggression (e.g., threatening, chasing), (2) contact aggression (e.g., biting, slapping), and (3) submission (e.g., grimacing, fleeing). The absolute frequencies of any single response, such as biting, were low and would require a considerably larger mass of data for meaningful analysis. It will be several more years before sufficient data accumulate. Table IV shows the number of responses per hour in the three broad categories listed, and analyses reveal some interesting relationships.

Average total agonistic rates were highest in the cynomolgus macaque group (12 per animal-hour). In day-to-day living, the gibbons showed little agonistic behavior of any description. The next lowest rates were found in the cebus group and one of the mangabey groups, both less than 4 per hour. The rate for the green monkeys was less than 5 per hour, whereas for all others, it varied between 6 and 8 per hour.

In the cebus and gelada groups submission accounted for the smallest fractions of agonistic behavior, 31 and 32%, respectively. The highest percentages were found in the rhesus, cynomolgus, and Celebes groups, 79, 74, and 69%, respectively. In the others, submission accounted for roughly half of all agonistic responses.

Total aggressive response rates were highest (above 3 per animal-hour) in the gelada, pigtailed macaque, cynomolgus, and one of the mangabey groups. The lowest rates, outside of the gibbon group, were seen in the rhesus and Celebes groups, with less than 2 occurrences per hour.

The distribution of aggressive responses between noncontact forms and those involving physical contact and potential physical damage reveals well-developed mechanisms ritualizing aggression and minimizing the risk due to intragroup fighting. In the green monkey group, where the average rate of involvement in aggressive interactions was relatively low (2 per hour), only 41% of aggressive scores were accounted for by noncontact responses. As a result, the consequences of aggression were often severe, despite the low frequency of aggression. In contrast, in the gelada group, where aggression was seen more frequently than in any other group (see Table IV), over 90% of aggressive scores comprised noncontact responses. In fact, despite the high overall rate of aggression, the geladas had the lowest scores for contact aggression. The rhesus groups also distributed their aggressive responses almost exclusively into noncontact forms, but these groups were generally inactive because of higher temperatures. The gibbons had near-zero agonistic totals, so information concerning the distribution of aggression between contact and noncontact forms is lacking, but in the few cases where aggression was seen, physical contact was involved. The only other group that channeled 50% or more of its aggression into contact aggression responses was the pigtailed macaques. This group, with relatively high aggressive response rates, had the highest contact aggression rates.

f. Sex. Another major response classification of great significance is sexual behavior and the responses associated with reproduction. For the purposes of this study all mounting responses and postures interpreted as soliciting mounting were combined in these two categories, i.e., no attempt was made to differentiate true reproductive behavior from similar-appearing response systems. From the data available, only the green monkey group showed any evidence for seasonal sexual behavior. In fact, examination of the birth records for all animals living outdoors at the Yerkes Field Station suggests a seasonal breeding peak only for the green monkeys; no sharp peak could be discerned from the records of the other groups. To be sure, gelada, Celebes, and mangabey records are still quite scanty, and mounting responses were undoubtedly incorporated into many social exchanges not related to reproduction (Bernstein, 1970b).

The rate of mounting behavior was, nonetheless, lowest in the gibbons, cebus, and green monkeys (Table IV). In all others, the hourly rate was one or two occurrences per animal-hour, although a rate in excess of 6 per hour had been previously observed in the pigtailed macaques. Postures interpreted as sexual solicitation were not seen in the three groups with the lowest mounting rates, and were rare in the cynomolgus and rhesus groups. Sexual solicitation rates were highest for the mangabeys (6 and 4 per animal-hour) and geladas (3 per hour).

The full sexual mount pattern was seen most frequently in the Celebes group, where one complete mount per animal-hour of observation was recorded. This was more than twice the rate seen in any other group. Full mounts were rarely seen in the green monkey group, and none was observed during formal observation sessions.

g. Lip-smacking. Lip-smacking was frequently associated with relatively friendly interactions, and almost invariably accompanied grooming. When scored only in nongrooming contexts, however, it was found to be most frequent in the Celebes group with an hourly frequency of 6 per animal (Table IV). A similar-appearing mouth movement in the Celebes group was recognized as a special response and was not included in these scores. The next highest rate for lip-smacking (4 per animal-hour) was found in the gelada group, followed by the cynomolgus group (2 per hour). The others showed hourly rates of 1 or 2, but lip-smacking was seldom seen in one of the mangabey groups, and never during a formal observation session.

h. Maternal care. Maternal responses such as ventral and dorsal carriage, the signals to initiate such carriage, and the support and restraint of infants and young juveniles, were found to relate so closely to age and number of infants present that no cross-taxa comparisons were attempted, inasmuch as insufficient age samples were available. Distinctive species responses could be noted, and relative distribution of dorsal to ventral carrying postures would easily distinguish certain taxa. The whole area of infant care is so rich, however (see Chapter 1 by Hinde), that a comparison of these responses must await more complete data collection and analysis.

i. Social huddling. One other social response deserves comment. Particularly during wet or cold weather, many animals may be seen to sit in a manner such as to expose a minimum of body area. In addition to mothers holding youngsters, unrelated animals often huddle together without clinging. These huddling postures involve considerable leaning of one animal against another and may also be seen during fair weather. During the fair-weather conditions of the present study, huddling was relatively rare, but reached highest expression in the geladas (4.5% of the time), a group also noted for the highest grooming and proximity scores, although nonspecific contact had been unremarkable. The pigtailed macaques had the next highest scores for huddling, about 3%, whereas little was seen in the cynomolgus macaques, and none in the mangabeys. In the earlier studies, the cebus monkeys were scored in huddles 8% of the time and the gibbons 6%, thus exceeding the gelada group scores. The rhesus macaques, with scores just under 3%, huddled almost as much as the pigtailed macaques.

2. INDIVIDUAL ACTIVITY PATTERNS

a. Inactivity. Periods of inactivity were combined for analysis regardless of whether the individual was resting, possibly asleep, or remaining quietly alert to his surroundings. The rhesus groups, scored at higher temperatures, spent 43% of the time inactive, whereas the next highest score was 19% for the green monkeys. The cynomolgus group had the lowest score (3.5%), and the gibbons spent only 7% of the time inactive. Scores for all other groups ranged from 11 to 16%. These data, summarized in Table III, are considered more accurate than the total group counts shown in Table II (see Section III, A).

b. Location. The two mangabey groups spent almost 33% of the sessions in their indoor quarters, the Celebes about 20%, and the other groups less than 10% of the time indoors (Table III). The geladas were almost always found outdoors. Although data-collection sessions were conducted only when temperatures ranged between 50 and 80°F, it must be noted that the geladas seldom came indoors during the daytime, even when temperatures were well below freezing. The larger-bodied forms, in general, could be found outdoors on any sunny winter day, even though temperatures were below freezing. Many animals also elected to sleep outdoors even when temperatures dipped to near 0°F.

Although all figures for location on the ground tend to be quite high under the particular conditions prevailing, it can be seen in Table II that the cynomolgus macaques, which adapt so well to life on the ground in botanical gardens, parks, and cities of southeast Asia, spent the most time on the ground. The ordinarily ground-living gelada baboons had the next highest scores, whereas the green monkeys made the most use of available trees. This conclusion is verified by the percentage of periods in which individuals spent some portion of the time in the highest locations or on the ground (Table III). The same relationships apply.

c. Demonstration displays. Vigorous physical demonstrations, such as branch-shaking, bouncing, or repeated banging of objects, are typical of many primates. These response patterns are clear and dramatic, but their interpretation is obscure. Certainly these activities attract attention to the performing individual, but the fact that so many animals appear to ignore such activity may mean that these are individual responses or that the social message is intended for animals outside of the immediate group. These performances are relatively rare in group-living animals, and although almost any individual may show the response pattern, they are most frequent in certain age–sex categories. During the present study, such demonstrations were seen in less than 1% of periods in the pigtailed macaque group, where they occurred most frequently. They were next most common in the Celebes group, and virtually absent in the green monkeys where several distinctive displays seemed to substitute in situations that elicited physical demonstrations in other groups.

d. Travel. Travel was considered as an individual activity and was scored any time an individual moved a straight-line distance in excess of one body length. As such, travel was a gross measure of general activity. Scores are summarized in Table III. The gibbons, which spent 73% of sessions traveling, had the highest scores. The next highest scores were 45 and 42% for cynomolgus and pigtailed macaques of the present study. The lowest scores were 19% for the geladas and 15% for the rhesus macaques. The discrepancy between the rhesus scores and those of the other macaques may be an effect of temperature.

e. Eating. In most groups, individuals were seen placing objects into the mouth (the operational definition of eating) in one-fifth to one-third of periods (Table III). The extreme low score was 6% in the rhesus group with the next lowest score, 16%, in the geladas. The high was scored in the cynomolgus group, 36%.

f. Object manipulation. Closely related to foraging and feeding was the manipulation of objects; scores are shown in Table III. Frequently, such manipulation could not be interpreted as related to feeding, but neither could it be separated from foraging patterns. The highest score for manipulation was seen in the pigtailed macaque group whose score rose from 4 to 12% in the new compound, where natural vegetation was available. Perhaps some undetermined difference in available stimuli also caused the difference between the 3 and 10% manipulation scores of the two mangabey groups. The gibbons were seldom if ever seen manipulating objects, and the low rhesus score may be attributed to factors already discussed. The green monkeys lived in a compound with a wealth of trees and brush and although they beat distinct pathways through the area, they did remarkably little damage to the vegetation. They were seen manipulating objects in only 2% of viewing periods, giving an impression of little interest in manipulable objects. The Celebes group, with a score of 8%, gave the impression of being the most manipulative group, and they earned a reputation for disassembling equipment and cages. Other impressions, only partially supported by quantitative data, are that the pigtailed and cynomolgus macaque groups show moderate interest in the manipulation of objects, whereas the cebus monkeys, with a reputation as skillful manipulators, do relatively little idle manipulation of objects. The geladas, with specialized grass-plucking and collecting movements, also have specialized digging movements (see Crook, 1966) and move objects about during foraging activities.

g. Self-directed activity. Self-directed manipulation often consisted of self-grooming. The cynomolgus macaques spent as much as 29% of the time so occupied, and it might be noted that this group had the second highest social grooming scores (Table III). Thus, self-directed grooming is not simply a substitute for grooming care received in social situations. Similarly, the geladas, with the second highest score for self-directed activity (28%), had the highest grooming scores. The gibbons and cebus monkeys, with the lowest social grooming scores, had the lowest scores for self-directed activities (9 and 11%, respectively). The rhesus monkey group and the pigtailed macaques, in their original living area, both spent about 12% of the time in self-directed activities, but in the new compound the pigtailed macaque score was higher and just barely came into the 19 to 25% range of the green monkey, mangabey, and Celebes groups. Thus, despite what seemed to be a sizable effect produced by moving the pigtailed macaque group to new quarters, their two scores did not bracket measures for any other group.

3. SPECIES-SPECIFIC PATTERNS

In addition to the time the groups spent in the broad activity categories just discussed, which were common to all taxa to some extent, specialized responses were also scored, each of which appeared in only one or two groups (see also Section III, D, 2). These specialized responses appeared as often as 4 per hour in some groups, e.g., the pigtailed macaques where the distinctive "pucker," consisting of the protrusion of pursed lips with the lower portion of the face thrust sharply forward, has already been reported (van Hooff, 1962; Kaufman & Rosenblum, 1966; Bobbitt, Jensen, & Gordon, 1964).

The Celebes black apes were seen to elevate the upper lip, thereby exposing the upper teeth and gums, at a rate of 2 per animal-hour. Similarly, specialized responses appeared with rates from 1 to over 4 per hour in all the other groups in the present study (see Table IV). If the activities of these groups were in no other way distinctive, the presence of the distinctive special responses would immediately distinguish individual response profiles as belonging to a particular taxon. This statement would also hold for the different green monkey taxa present at the Yerkes Field Station as members of other study groups.

4. DISCUSSION OF RESPONSE PROFILES

The range of variation in activity patterns and the resulting profiles suggest that each taxon may be represented by a distinctive activity profile. Although variables such as age, sex, status, and environmental factors (e.g., temperature and physical features), all may be demonstrated to influence the activity profiles, these influences may vary from taxon to taxon and should not completely mask taxonomic effects. Nevertheless, these factors must be controlled if systematic changes in activity profiles are to demonstrate the degree of affinity between related taxa, and succeed in identifying congeneric forms as more similar than taxa separated at the family level. No single activity measure can be expected invariably to demonstrate this, but the data of Davis *et al.* (1968) suggest that the whole activity profile may do so. Because of the number of important low-frequency items, a considerable mass of data may be required and the effort can be justified only if it results in a system which can identify the species of each individual and the relationships between taxa. Relationships among activity profiles may parallel those based on phylogeny, ecology, morphology, or some similar factor or combination of factors.

The primate taxa examined in the program just reported can be differentiated by their activity profiles albeit only relatively gross response classifications were considered. For example, the gibbon profile is recognizable by the highest recorded score for travel coupled with the lowest recorded score for object manipulation. The cebus monkeys and gibbons are differentiated from the Old World monkeys (Cercopithecoidea) by their high nonspecific contact scores coupled with low grooming scores. On the other hand, the gibbons are distinct from any of the monkey groups in their high play scores and nearly absent agonistic responses. The green monkeys are distinct among the Old World monkeys studied because of their extremely low sexual response rates, the high percentage of contact aggression compared to noncontact aggression, and their low grooming scores. The mangabeys, cebus monkeys, and geladas could be differentiated from the others by low social-play scores, but whereas the cebus monkeys showed few submissive responses and the geladas relatively few submissive responses with a great preponderance of noncontact aggression, and whereas cebus monkeys showed few sexual responses, the mangabeys tended to be intermediate in all these categories, as well as in most others. The geladas had the highest proximity and grooming scores as well as relatively low travel scores, which were not coupled with inactivity.

Thus, all the genera could be recognized as distinct, and the greater taxonomic separation of capuchins and gibbons could be recognized. It should also be noted that these groups are morphologically distinct with the least variation occurring among the Old World monkeys. Gibbons are apes; they are also brachiators and live in monogamous families. Cebus monkeys are New World monkeys; they also have small body size, many morphological adaptations, and are highly arboreal. Geladas are perhaps the most completely adapted to terrestrial living as well as having the largest body size and characteristically living in one-male units within a troop structure. The single guenon *(Cercopithecus)* group studied is of some interest in that it is partially terrestrial (although the green monkeys made more use of the trees than did the geladas, Celebes, mangabey, or macaque groups), is of intermediate body size (the West African form is the largest green monkey taxon, and shows clear sexual dimorphism), and was still easily differentiated from the other genera of Old World monkeys.

Among the macaque and Celebes groups, the data for the rhesus macaque are aberrant, but were strongly influenced by high temperatures. To be sure, levels of sexual response, the high level of submissive behavior and relatively low level of contact aggression were similar for the rhesus and cynomolgus groups. The extremely high score for inactivity, on the other hand, was clearly a temperature effect and should be compared with the extremely low score in the cynomolgus group. Grooming, contact and proximity scores seem unremarkable, but the general picture of extreme inactivity coupled with low levels of feeding, travel, and play, and relatively little object manipulation and even self-directed activity all seem related to the single variable, high temperature.

The activity profile for the Celebes group could not be readily differentiated from those of the macaques on the basis of the broader categories discussed. Certain special responses were distinctive, but special responses could also be used to identify individual macaque species. A distinctive combination within the profile did exist for the Celebes group but such combinations could be found for all groups, and it is possible to distinguish any of the macaque species from the rest of the genus with the same ease as it is to distinguish the Celebes groups with the data presented. As a matter of fact, the influence of variables such as body size might account for our ability to distinguish congeneric forms, but we can also readily detect similarities within a genus and yet differentiate genera within the Old World monkeys, so systematic variables apparently are operating. Thus the Old World monkeys, in general, share certain similarities as compared with the grossly different and taxonomically distinct cebus monkeys and gibbons.

The Celebes group could be immediately recognized as distinct from the other six groups of the present study by examining the data for lip-smacking. As seen in Table IV, these data are not available for the earlier study groups but the Celebes scores are higher than those for any other taxon studied, including the geladas whose scores are higher than those of any macaque group. This relationship applied even when a distinctive lip-movement, peculiar to the Celebes and similar to lip-smacking, was not included in the scoring.

It might also be noted that although sexual activity was not remarkably high in the Celebes, the full mounting pattern was seen more frequently in this group than in any other group. Thus, the Celebes group, whose taxonomic status remains unclear, could be differentiated from the macaques, but only by including more information than required for the other genera.

D. Sex Differences

In addition to taxonomic comparisons, the data collected were used to investigate male and female role effects as expressed in activity-profile differences. Since precise sexual composition of a group may be characteristic of a primate taxon, it was not controlled in this study and could conceivably bias the results obtained. The following analysis demonstrates the effect of sex roles in all of the groups for which data were obtained. The data are summarized in Tables V and VI, which are derived from the average scores for males and for females in each of the groups.

1. SOCIAL ACTIVITIES

a. Maternal responses. Maternal behavior is one of the first areas in which clear sex differences in active response participation would be expected (see Chapter 1 by Hinde). Indeed, in the taxa included in the present study, females carried and cared for young most intensively, and this behavior was seen not only in the biological mothers of infants, but also in immature females. These patterns were not exclusively female activities, however, and male cynomolgus and pigtailed macaques, green monkeys, and geladas were all seen to carry infants and young animals. The adult bachelor male gelada had a very special role in that he frequently played with, carried, and otherwise interacted with infants, in prolonged intensive interaction sequences.

b. Proximity and contact. Except for one mangabey group, where males spent more time in nonspecific contact, females tended to spend more time in proximity to and in nonspecific contact with others (Table V).

c. Total social responses. Gross differences between male and female total social-interaction scores were investigated by pooling all scores related to social behavior. In every group, both sexes initiated more social interactions than they received. This apparent incongruity can be accounted for when it is understood that several animals may chase, groom, make sexual overtures to, or sniff a single animal. In some cases, the recipient of a social message could not be precisely identified and hence only the actor was scored. The net effect was that higher scores for participation in the role of actor than in that of recipient were found in both sexes. Little sexual difference could be seen with the possible exception of the green monkeys, where females appeared to be more frequently involved in social interactions (Table VI). The percentage of *time* individuals made and received social responses did differentiate the two sexes, despite the nearly equal *frequencies* of initiation and receipt. In every group, the males spent more time

TABLE V

Male–Female Ratio on Percentage of 30-Second Periods in Selected Activities and Locations

Activity or Location	Pigtails	Cynomolgus	Celebes	Geladas	Mangabeys 1	Mangabeys 2	Greens
Proximity to others	0.9	1.0	0.9	0.8	0.7	0.8	0.8
Nonspecific contact	0.6	0.8	0.4	0.7	0.4	1.2	0.8
Grooms others	0.4	0.3	0.9	0.2	0.6	0.1	0.3
Receives grooming	1.0	0.7	1.0	0.6	0.7	0.8	0.3
Play	1.7	9.+	0.5	0.2	0.7	–	2.2
Inactive	1.1	1.3	1.7	1.7	1.9	0.8	2.4
Highest locations	3.0	1.1	0.7	2.6	4.4	–	0.–
On the ground	1.0	1.0	1.0	0.9	0.9	1.3	0.9
Demonstration displays	1.6	4.0	3.0	9.+	0.–	–	–
Object manipulation	1.2	1.2	0.7	1.2	1.1	0.2	4.1
Self-directed activity	0.9	1.0	1.4	1.5	1.5	0.9	1.5
Travel	0.9	1.5	0.7	0.8	0.9	1.5	1.0
Eating	0.9	1.6	0.6	0.5	1.0	2.9	0.9

NOTE: Ratios in excess of 1.0 indicate higher males scores, e.g., 4.0 means males were involved four times more often than females. Score of 9.+ indicates a ratio greater than 10; a score of 0.– indicates a ratio less than 0.05. A – means no episodes were scored.

TABLE VI

Male–Female Ratio on Hourly Frequency of Participation in Selected Activities

Activity	Pigtails	Cynomolgus	Celebes	Geladas	Mangabeys 1	Mangabeys 2	Greens
Noncontact aggression (P)	0.6	1.2	0.5	0.5	1.0	6.3	0.9
Noncontact aggression (R)	0.6	1.5	0.1	0.8	0.1	0.3	0.4
Contact aggression (P)	0.1	1.1	0.2	0.–	1.6	9.+	1.2
Contact aggression (R)	1.2	6.3	0.3	1.0	0.–	0.–	0.3
Total aggression (P)	0.4	1.1	0.3	0.4	1.1	7.5	1.1
Total aggression (R)	1.0	1.7	0.2	0.9	0.–	0.2	0.3
Submission (P)	0.6	9.+	0.5	0.1	9.+	1.3	0.3
Submission (R)	0.5	1.3	0.8	1.1	6.4	4.2	0.1
Total agonistic	0.6	1.3	0.5	0.5	1.5	2.0	0.4
Mounting (P)	3.7	9.+	8.4	3.0	3.9	9.+	9.+
Mounting (R)	0.7	0.4	0.5	0.–	0.–	0.–	–
Sexual soliciting (P)	0.2	0.–	0.3	0.1	0.3	0.–	–
Sexual soliciting (R)	9.+	9.+	9.+	2.5	4.6	9.+	–
Genital investigation (P)	2.0	6.0	1.8	3.0	0.9	5.0	9.+
Genital investigation (R)	0.6	0.4	0.7	0.4	1.0	0.–	–
Lip-smacking (P)	0.5	0.5	0.6	1.4	7.5	–	0.4
Lip-smacking (R)	6.5	0.4	0.5	0.7	2.4	–	0.1
Species-specific patterns (P)	1.9	1.0	1.5	1.5	1.1	2.1	9.+
Species-specific patterns (R)	1.4	1.0	0.8	1.4	–	0.–	2.0
Total social responses (P)	1.0	1.4	1.0	1.1	0.9	2.4	0.7
Total social responses (R)	1.0	0.9	0.6	1.0	1.6	1.8	0.4

NOTE: Ratios in excess of 1.0 indicate higher male scores, e.g., 4.0 means males were involved four times more often than females. A score of 9.+ indicates a ratio greater than 10; a score of 0.– indicates a ratio less than 0.05. A – means no episodes were scored. (P) indicates that an animal performs the activity; (R) indicates that an animal is the recipient of the activity.

receiving than they did performing and always spent less time as actors than did
the females. The males also spent less total time participating in social
interactions than did the females, with the possible exception of the Celebes
group, where total interaction times were approximately equal. Female patterns
were generally the inverse of male patterns, and they spent more time as actors
than as recipients, with the possible exception of the mangabey groups, where
the two scores were nearly equal.

   d. Grooming. What produces these gross differences in male and female
interaction patterns? Perhaps the single social interaction accounting for the
greatest percentage of social interaction time accounts for much of these
differences. In every group, the females spent more time grooming than did
males, but they groomed both males and females such that females also spent
more time being groomed than did males, except in the Celebes and pigtailed
macaque groups, where the two scores were nearly equal (Table V). Sex
differences in participation in social grooming were also reported by Rosenblum,
Kaufman, and Stynes (1966) for bonnet and pigtailed macaque groups.

   Except in the Celebes and green monkey groups, where nearly equal scores
were recorded, males spent less time grooming than being groomed, whereas
females' scores were all nearer to equality with a tendency to groom more than
to be groomed. Presenting for grooming also seemed more common in females,
but some variation existed. The interrelationships of grooming, mounting, and
agonistic patterns in these groups have been analyzed by Bernstein (1970b).

   e. Genital investigation. Genital investigation (defined as always including
manipulation of the genitalia) was scored separately from grooming, and here
male and female sexual roles seemed to determine the directionality of
responses. As shown in Table VI, males generally inspected and females were
inspected, except in one mangabey group. In that group, many females also
inspected females, and in both the mangabey and Celebes groups many males
inspected males. These data were partially confounded by the frequency with
which infants of both sexes received genital investigations, but the adult sex
roles overrode these effects.

   f. Sexual responses. Primary sexual behavior was also determined essentially
by sex roles, with females and males presenting and mounting as appropriate in
the majority of instances (Table VI). Olfactory examination of others (social
sniffing, Table IV) was done primarily by males, but the extent to which this is
related to obtaining sexual cues is unknown. Certainly, sniffing of the mouth
area is not likely to be related to sexual cues, but much of the sniffing was
directed to the trunk and genitalia. The effectiveness of sexual pheromones has
been experimentally demonstrated with rhesus monkeys (Michael & Keverne,
1968).

   g. Agonistic responses. Agonistic behavior was not as clearly divided by
sexual membership and, as seen in Table VI, only in the cynomolgus macaques

and mangabeys were males more frequently involved than females. In these two groups, and in the green monkeys, males were more often involved in aggression, although one of the mangabey groups appears contradictory because of the high frequency with which females received aggression. In all groups except the cynomolgus macaques, females received at least as much aggression as did males. Again except for the cynomolgus macaques and mangabeys, females showed submissive responses more frequently than did males, a direct correlation with receipt of aggression, perhaps. In the two exceptional groups, and the geladas, the males received more submissive responses than did the females.

The females in the Celebes, and pigtailed macaque groups had higher frequency scores than the males for initiating each of the responses classified as aggression. In the other groups, males did more of the damaging contact aggression (biting primarily) than did the females, whereas females did more of the slapping and pulling, except in the mangabeys and the cynomolgus macaques, where the scores were nearly equal. Green monkey males did more threatening and chasing than did the females. The males in one mangabey group also did more threatening than did the females. In the other groups, the females appeared to do most of the threatening.

The males were more frequently targets of contact aggression only among the pigtailed and cynomolgus macaques. Among the geladas, both sexes received contact aggression equally frequently. The females in all groups except the cynomolgus macaques also received more noncontact aggression. In every group except this group, the males showed more noncontact aggression than they received.

h. Play. Play was more common in the male pigtailed and cynomolgus macaques and male green monkeys, but was more common in females in the other groups (see Table V). This is perhaps related to age composition, but it might be noted that the adult gelada females had high play scores. All four of the immature animals in this group were males, but the females still had higher play-participation scores.

i. Lip-smacking. Among the mangabeys, the majority of lip-smacking responses were both directed toward the males and were made by them. In the gelada group, the males made more and in the pigtailed macaques more were directed toward the males. In the other groups, females made more lip-smacking responses and more were directed toward them (see Table VI). This may be due in part to the higher proximity scores and approaches by females to females with young infants. Lip-smacking was also performed by approaching animals in many other nonhostile situations, and seems to serve a general reassurance function.

2. SPECIES-SPECIFIC PATTERNS

Many of the specialized responses characteristic of a single species were performed almost exclusively by males (see Table VI). Thus, pigtailed macaque males do most of the puckering, the combination of lip-smacking and tongue-protrusion seen in mangabeys is done mostly by males to females, and the high grin of the Celebes black apes is seen slightly more often in males. Tail-arching, repeated sitting upright, and various weaving and bouncing sequences in the green monkeys were also primarily male responses. In all groups, these responses were seen when individuals approached or were approached by another. The subsequent response patterns tended to be aggressive in the green monkeys, sexual in the mangabeys, variable in the pigtailed macaques, and nonaggressive in the Celebes black apes. (The Celebes or moor macaque does not show this response, and adults of that species show certain other distinctive patterns not seen in the Celebes black apes.) Tail-arching in sooty mangabeys seems to be a general alerting posture and was seen equally often in both sexes. A number of types of specialized gelada responses were scored, and males were most frequently the actors and the recipients. However, contrary to the relationships seen in the other taxa, most types of specialized responses were characteristic of females, often involving the female "bead" patterns and their display. Examination was frequently but not exclusively done by males. The eversion of the upper lip over the nostrils, which is such a striking gelada expression, was done more often by females.

3. Individual Activities

Males spent more time inactive than females in every group except one of the mangabey groups (see Table V), primarily because of the male habit of sitting passively alert. Such alert postures were often assumed in the top of the highest available tree so that, with the exception of the green monkey and Celebes groups, males tended to spend more time in the highest locations than did females. Violent physical demonstrations such as branch-shaking were also primarily male activities, and perhaps related to the alertly watchful attitudes, but such demonstrations were also occasionally performed by females.

Object manipulation was also more characteristic of males, except in the Celebes and one mangabey group. This perhaps relates to the exploratory tendencies of juvenile and adolescent males in particular, as for example reported in the striking preponderance of males among animals exploring novel objects presented to a group of pigtailed macaques (Bernstein, 1966). In the gelada, Celebes, green monkey, and one of the mangabey groups, the males also showed more self-directed activity; in the other groups, their scores were similar to the female scores. This higher self-directed activity score could be related to the generally lower participation in social grooming characteristic of males.

Male and female scores seemed variable or equal in such categories as time spent indoors, travel, feeding, and related activities.

4. MALE AND FEMALE ROLES AND ACTIVITY PROFILES

In general, male and female role patterns can be identified in differential activity profiles, portions of which were common to all six primate taxa. As should not be surprising, sex-determined role functions are not identical in all primate taxa, despite certain broad similarities. The different sex-related activity profiles may help to identify differences in sex roles in the primate taxa under study.

IV. GENERAL DISCUSSION

In the study of primates living as members of a society, it is the organized group that is the subject, and each group must be considered as an individual subject. Both the physical and behavioral aspects of the society must be described and measured. In the case of the group, the physical characteristics include not only such obvious features as age and sex composition and total size, but also such aspects as the longevity of associations and genealogical relationships. The social structure and social mechanisms are undoubtedly intertwined with these latter considerations, but it is my view that behavior and morphology are always so interrelated.

Furthermore, much as the immediate stimulus environment may influence behavior and produce learning, so may long-term environmental influences affect behavior. That is to say that the selective pressures operating in the natural habitat of a species will determine the particular form of response or response patterns. An animal, therefore, comes into any situation with a limited number of possible responses available to it, and some of these responses may be innately prepotent and more likely to be elicited in any particular situation. Whereas morphology alone may limit possible facial expressions, body postures, and motor patterns, the actual response repertoire is further restricted by long-term variables that have biologically "reinforced" or "rewarded" certain types of responses. This is just another way to view long-term selective pressures and their results as parallels to short-term individual learning (cf. Skinner, 1966). Genetic encoding may be viewed as species modification much as individual behavior modification is viewed as learning. The apparent increase in learning potential to be found in the order Primates must result from some process that has selected for increased plasticity of behavior in primates. We can only speculate that increasing the degree of behavior modification possible must, in some way, be adaptive in the particular ecological niches exploited by primates. Despite this increase in learning potential, we cannot ignore the underlying innate bases for most behavior. Although both the form and frequency of a behavior pattern may be modifiable in response to a number of environmental variables, the innate repertoire will strongly influence the probability of certain patterns and their precise form of expression.

Thus, in a specified controlled environment, an organized primate group should show consistent behavior patterns. These group patterns should be expected to vary in much the same manner as individual responses are found to vary. Minor changes in the environment will influence the resulting group behavior pattern and may be viewed much as the experimental manipulation of environmental variables in studies of individual behavior. Interindividual variation will find its parallel in the variation to the expected between groups of the same type of animal. Age, sex, and size variables in individuals may find expression in the social group by manipulation of these factors in the composition of the groups. The greater the degree of variation between two subject populations, the easier it should be to distinguish them. In the study of organized groups, widely disparate populations can be obtained through a comparative approach using different primate species. Such comparative studies of social groups should reveal broad ranges of differential behavior, both with respect to response frequencies and with respect to response expression.

Each primate species is postulated to have a species-typical response repertoire with certain responses possibly serving as species-specific indicators. Furthermore, morphological and phylogenetic differences among primate taxa will produce differences in the frequencies and durations of broad classes of activities, resulting in distinctive activity profiles for each primate taxon. The degree of difference among activity profiles should reflect the phylogenetic relationships of the animals compared, although differences and similarities reflecting habitats and responses to specific experimental variables may also influence the resulting activity profiles.

These considerations are often the rationale behind behavioral field studies. Responses are described within the presumed habitat in which they evolved. The same responses can, of course, be studied under less relevant but more specifically controlled conditions. The comparison of species' responses and the ability of a taxon to modify its responses is the subject matter of comparative psychology.

This chapter has presented the results of a comparative approach that demonstrates the feasibility of such a procedure in the study of social groups. Despite the difficulties in controlling a number of variables (a constant problem in comparative psychology at any level), the results demonstrate that the frequencies and durations of the activities of a social group can be used as an indicator of the phylogenetic position of that group. Additional data on responses to variables such as temperature and the physical features of the test situation serve to indicate that particular responses may be heavily influenced by such variables, but it is the activity profile as a whole that is indicative. The influence of such variables upon the activity profile is assumed to be typical of the taxon under study, and the changes in activity profile produced by manipulation of such factors may themselves be useful in identifying special adaptive features of the behavior of a species.

In the work presented in this chapter, data on six primate taxa were collected under similar test conditions, in order to provide a test of these hypotheses. One taxon was represented by two groups (the sooty mangabeys) and another was tested under somewhat different conditions as well as under the standard test conditions (the pigtailed macaques). In addition, data for three other primate taxa were included for comparison purposes (gibbons, cebus monkeys, and rhesus monkeys). The primates selected for study thus included three congeneric forms (rhesus, pigtailed, and cynomolgus macaques), another closed related form (the Celebes black apes), three other genera of Old World monkeys (the green monkeys, geladas, and sooty mangabeys), one New World monkey (the cebus monkeys), and one ape (the gibbons).

Reviewing the data collected against a background of ecological habitat descriptions and phylogenetic classifications reveals activity profiles that are indicative of phylogeny, whereas specific activity measures may reflect ecological adaptations. For example, although the total activity profiles clearly distinguished the macaques from the other taxa and indicated strong relationships among macaque activity profiles, the selection of a single measure such as location on the ground would have shown the geladas to be intermediate between the pigtailed and cynomolgus macaques, whereas the Celebes black apes, which are closely related to the macaques, show scores similar to those of· the mangabey groups (see Table II). Such scores probably reflect the readiness with which these species descend to the ground in the natural habitat. The cynomolgus macaques quickly invade botanical gardens and clearings, whereas the pigtailed macaques seldom leave the forest. The gelada is well adapted to life on the ground and the mangabeys are essentially arboreal forest forms. Although no information is available for the Celebes black apes, the data would suggest these are primarily forest forms, perhaps similar to the pigtailed macaques. Figures for the green monkeys suggest a primarily arboreal adaptation despite the propensity for coming to the ground seen in this one species of guenon.

Reading across any single line in Table III might suggest other spurious relationships between taxa. Cebus monkeys and gibbons, two widely disparate primates, show similar scores for proximity, grooming, nonspecific contact, and self-directed activities. These could be simple reflections of their arboreal adaptations, for the wide disparities in scores for travel and play and differences between these two groups in other scores in Tables II, III, and IV preclude the similarities predicted for phylogenetically closely related forms. The pigtailed and cynomolgus macaques, in contrast, show basically similar profiles despite differences in a few categories. The few differences can easily be explained as species specializations related to primary habitats and gross body size. Even in some categories useful in discriminating different species of macaques, the macaque scores are relatively similar when compared with the scores of groups from other genera. Even the rhesus monkey scores collected under conditions of high temperature (which has been demonstrated to produce a profound effect on primate activities) produce a total profile compatible with the other macaques. The Celebes group, with a problematic phylogenetic position, clearly show close affinities to the macaques, although a few specialized responses can be found to demonstrate their distinctive position.

Data analysis in this study did not go beyond determining the frequency and duration of selected responses and activity patterns. An analysis of response sequences would be required before social mechanisms and role patterns could be identified. Nevertheless, we can speculate on the relationship of the selective forces in the natural habitat and the activity patterns as measured.

The data for location on the ground and in the highest available locations (Tables II and III) strongly suggest the arboreal adaptation of the green monkeys. This adaptation is typical of the guenons, and the green monkeys must be considered to have only recently begun to exploit the resources of their present habitat along the fringes of the savannah. The relationships among arboreal modes of life and other activities seem unclear, however, in that the gibbons and cebus monkeys are among the most arboreal and the geladas the least, yet few activities would reveal this scale of adaptation. In fact, the brachiating locomotor specialty of the gibbons, and their distinct phylogenetic position, probably would completely overshadow any generalized arboreal effect.

Proximity scores underline the fact that all of the primates under study are social creatures. Most individuals spend between 50 and 75% of the time within a body length of another, despite ample opportunities for greater scatter. Nonspecific contact scores were much more variable, indicating strong contact motivation among the gibbons and cebus monkeys and far less among the mangabeys and pigtailed macaques. This is not immediately explainable as a response to agonistic patterns or sexual behavior (variable scores in Table IV), nor can it be readily attributed to the extent of maternal relationships, which was greatest for the pigtailed and cynomolgus macaque groups. Even phylogenetic relationships would be inadequate inasmuch as the pigtailed macaque scores do not parallel those for the other macaques, and the cebus monkeys and gibbons are very disparate forms. Closer examination of the ecological niches and special adaptations may suggest why mangabeys and pigtailed macaques have similar scores.

Travel scores are highest in the brachiating gibbons and lowest in the grazing geladas, which spend much of the day sitting, including feeding times during which they shuttle along plucking grass rather than walking from place to place.

Several different mechanisms seem to control the consequences of agonistic encounters. A group such as the cynomolgus macaques relied heavily on submissive signals, whereas the geladas channeled most aggression into noncontact responses (Table IV). The presence or absence of a strict dominance hierarchy could not be detected in such scores; the determination of social status would require an analysis of sequences between dyads, and this was not attempted with the data presented. It might be noted that other data suggest little evidence for linear social-status relationships among cebus monkeys, gibbons, or green monkeys (Bernstein, 1970b). Whereas the total agonistic scores for these animals were low but not truly remarkable, these groups also showed the lowest scores for sexual responses. The interrelationships between dominance, the expression of aggression, and its influence on the signal function of sexual responses still remain to be demonstrated, but the present data do not preclude such relationships. The exact nature of interrelationships, however, cannot be specified from frequency data alone.

The data presented in Tables V and VI describe behavioral comparisons between the sexes in the species under consideration. Sexual dimorphism exists in all the groups studied, and is perhaps greatest in the geladas. The degree of physical differentiation of the sexes is not paralleled in the degree of distinctive behavioral adaptations for the two sexes. In fact, the differences in behavior which can be attributed to sex more likely correlate with social roles in the different types of social organizations represented. Thus, the geladas base their societies on associated one-male groups whereas more than one adult male and a pool of females were common to the other groups. (Gibbon data were not available.) Infant care patterns and social status hierarchies, or their absence, can account for much of the differences in degree of sex-related activity patterns.

What is clear in Tables V and VI is that, once again, distinctive species profiles can be demonstrated. Thus, the measures of frequencies and durations of activity patterns can be used readily to differentiate species-typical activity profiles. Furthermore, activity profiles have been demonstrated to be sensitive to changes in a large number of variables. These activity profiles, however, are themselves insufficient in the study of social organization. The same data must also be analyzed for sequential patterns in order to identify roles, social mechanisms, and the interrelationships among individuals within an organized social group. We cannot short-circuit the four necessary steps of (1) response identification, (2) frequency and duration measurement, (3) sequential analyses, and (4) synthesis into the organizing principles of the society under study.

V. SUMMATION

The purpose of this chapter has been to present a series of quantitative measures of social and individual activities whose total patterns could be used to differentiate primate taxa. Taxonomic affiliation was seen to produce recognizably different activity profiles in the six primate taxa studied in groups maintained under controlled conditions. Three other taxa maintained under different captive group conditions could also be differentiated. Multiple measures on the same primate taxa tended to produce similar profiles, although specific environmental variables produced sizable systematic effects on the frequencies and durations of responses. Sex differences in all taxa were demonstrated, but sex effects were not identical in all species. Age effects were assumed to be large, but quantitative data demonstrating these influences will await larger samples including longitudinal data on the same individuals.

The activity profiles are suggested to be sensitive to individual factors such as age, size, and condition, and to environmental influences such as temperature, time of day, weather, and physical spatial features. Phylogenetic influences are nonetheless discernible, and congeneric forms can be differentiated while widely divergent forms can be identified as separated by more basic differences in activity profiles.

The combined influences of ecology, morphology, and phylogeny thus result in distinctive activity profiles for primate social groups. By carefully controlling or measuring the influences of these variables, the influence of any single variable may be determined. This procedure is offered as a tool in the comparative study of primate behavior. It should be regarded as a supplement to and not a substitute for other techniques that have already proven their value. The data presented in this chapter represent a beginning.

REFERENCES

Altmann, S. A. The social behavior of anthropoid primates: An analysis of some recent concepts. *In* E. L. Bliss (ed.), *Roots of behavior.* New York: Harper & Row, 1962. Pp. 277-285.

Bernstein, I. S. A field study of the activities of howler monkeys. *Animal Behaviour,* 1964, **12**, 92-97. (a)

Bernstein, I. S. The integration of rhesus monkeys introduced to a group. *Folia Primatologica,* 1964, **2**, 50-63. (b)

Bernstein, I. S. Activity patterns in a cebus monkey group. *Folia Primatologica,* 1965, **3**, 211-224.

Bernstein, I. S. An investigation of the organization of pigtail monkey groups through the use of challenges. *Primates,* 1966, 7, 471-480.

Bernstein, I. S. A field study of the pigtail monkey *(Macaca nemestrina). Primates,* 1967, **8**, 217-228. (a)

Bernstein, I. S. Defining the natural habitat. *In* D. Starck, R. Schneider, & H.-J. Kuhn (eds), *Progress in primatology.* Stuttgart: Fischer, 1967. Pp. 177-179. (b)

Bernstein, I. S. The lutong of Kuala Selangor. *Behaviour,* 1968, **32**, 1-16.

Bernstein, I. S. Introductory techniques in the formation of pigtail monkey troops. *Folia Primatologica,* 1969, **10**, 1-19. (a)

Bernstein, I. S. Stability of the status hierarchy in a pigtail monkey *(Macaca nemestrina)* group. *Animal Behaviour,* 1969, **17**, 452-458. (b)

Bernstein, I. S. Activity patterns of pigtail monkey groups. *Folia Primatologica,* 1970, **12**, 187-198. (a)

Bernstein, I. S. Primate status hierarchies. *In* L. A. Rosenblum (ed.), *Primate behavior.* Vol. 1. New York: Academic Press, 1970. Pp. 71-109. (b)

Bernstein, I. S. Some behavioral elements in the Cercopithecoidea. *In* J. R. Napier (ed.), *The Old World monkeys.* New York: Karger, 1970. Pp. 265-295. (c)

Bernstein, I. S., & Draper, W. A. The behaviour of juvenile rhesus monkeys in groups. *Animal Behaviour,* 1964, **12,** 84-91.

Bernstein, I. S., & Mason, W. A. Activity patterns of rhesus monkeys in a social group. *Animal Behaviour,* 1963, **11,** 455-460.

Bernstein, I. S., & Schusterman, R. J. The activities of gibbons in a social group. *Folia Primatologica,* 1964, **2,** 161-170.

Bernstein, I. S., & Sharpe, L. G. Social roles in a rhesus monkey group. *Behaviour,* 1965, **26,** 91-104.

Bobbitt, Ruth, A., Jensen, G. D., & Gordon, Betty N. Behavioral elements (taxonomy) for observing monther-infant-peer interaction in *Macaca nemestrina. Primates,* 1964, **5,** 71-80.

Bourlière, F. Patterns of social grouping among wild primates. *In* S. L. Washburn (ed.), *Social life of early man.* Chicago: Aldine, 1961. Pp. 1-10.

Carpenter, C. R. A field study of the behavior and social relations of howling monkeys *(Alouatta palliata). Comparative Psychology Monographs,* 1934, **10,** No. 2 (Whole No. 48).

Carpenter, C. R. A field study in Siam of the behavior and social relations of the gibbon *(Hylobates lar). Comparative Psychology Monographs,* 1940, **16,** No. 5 (Whole No. 84).

Collias, N., & Southwick, C. A field study of population density and social organization in howling monkeys. *Proceedings of the American Philosophical Society,* 1952, **96,** 143-156.

Crook, J. H. Gelada baboon herd structure and movement: A comparative report. *Symposia of the Zoological Society of London,* 1966, **18,** 237-258.

Crook, J. H. Evolutionary change in primate societies. *Science Journal,* 1967, **3,** 66-70.

Davis, R. T., Leary, R. W., Casebeer Smith, Mary D., & Thompson, R. F. Species differences in the gross behaviour of nonhuman primates. *Behaviour,* 1968, **31,** 326-338.

Eisenberg, J. F., & Kuehn, R. E. The behavior of *Ateles geoffroyii* and related species. *Smithsonian Miscellaneous Collections,* 1966, **151,** 1-63.

Ellefson, J. O. Territorial behavior in the common white-handed gibbon, *Hylobates lar* Linn. *In* Phyllis C. Jay (ed.), *Primates.* New York: Holt, Rinehart and Winston, 1968. Pp. 180-199.

Gartlan, J. S., & Brain, C. K. Ecology and social variability in *Cercopithecus aethiops* and *C. mitis. In* Phyllis C. Jay (ed.), *Primates.* New York: Holt, Rinehart and Winston, 1968. Pp. 253-292.

Hall, K. R. L. Variations in the ecology of the chacma baboon, *Papio ursinus. Symposia of the Zoological Society of London,* 1963, **10,** 1-28.

Hall, K. R. L. Behaviour and ecology of the wild patas monkey, *Erythrocebus patas,* in Uganda. *Journal of Zoology,* 1965, **148,** 15-87.

Hall, K. R. L., & DeVore, I. Baboon social behavior. *In* I. DeVore (ed.), *Primate behavior.* New York: Holt, Rinehart and Winston, 1965. Pp. 53-110.

Hooff, J. A. R. A. M. van. Facial expressions in higher primates. *Symposia of the Zoological Society of London,* 1962, **8,** 97-125.

Kaufman, I. C., & Rosenblum, L. A. A behavioral taxonomy for *Macaca nemestrina* and *Macaca radiata. Primates,* 1966, **7,** 205-258.

Kummer, H. *Social organization of hamadryas baboons.* Chicago: Univ. Chicago Press, 1968.

Kummer, H., & Kurt, F. A comparison of social behavior in captive and wild hamadryas baboons. *In* H. Vagtborg (ed.), *The baboon in medical research.* Austin: Univ. Texas Press, 1965. Pp. 65-80.

Lawick-Goodall, Jane van. A preliminary report on expressive movements and communication in the Gombe Stream chimpanzees. *In* Phyllis C. Jay (ed.), *Primates.* New York: Holt, Rinehart and Winston, 1968. Pp. 313-374.

Loizos, Caroline. Play behavior in higher primates: A review. *In* D. Morris (ed.), *Primate ethology.* London: Weidenfeld and Nicolson, 1967. Pp. 176-218.

Mason, W. A. Social organization of the South American monkey, *Callicebus moloch:* A preliminary report. *Tulane Studies in Zoology,* 1966, **13,** 23-28.

Mason, W. A. Use of space by *Callicebus* groups. *In* Phyllis C. Jay (ed.), *Primates.* New York: Holt, Rinehart and Winston, 1968. Pp. 200-216.

Michael, R. P., & Keverne, E. B. Pheromones in the communication of sexual status in primates. *Nature,* 1968, **218,** 746-749.

Poirier, F. E. The Nilgiri langur *(Presbytis johnii)* troop: Its composition, structure, function and change. *Folia Primatologica,* 1969, **10,** 20-47.

Reynolds, V. An outline of the behavior and social organization of forest living chimpanzees. *Folia Primatologica,* 1963, **1,** 95-102.

Reynolds, V., & Reynolds, Frances. Chimpanzees of the Budongo Forest. *In* I. DeVore (ed.), *Primate behavior.* New York: Holt, Rinehart and Winston, 1965. Pp. 368-424.

Rosenblum, L. A., Kaufman, I. C., & Stynes, A. J. Individual distance in two species of macaque. *Animal Behaviour,* 1964, **12,** 338-342.

Rosenblum, L. A., Kaufman, I. C., & Stynes, A. J. Some characteristics of adult social and autogrooming patterns in two species of macaque. *Folia Primatologica,* 1966, **4,** 438-451.

Rowell, T. E. A quantitative comparison of the behaviour of a wild and a caged baboon group. *Animal Behaviour,* 1967, **15,** 499-509.

Sarbin, T. R. Role theory. *In* G. Lindzey (ed.), *Handbook of social psychology.* Vol. I. Reading, Mass.: Addison-Wesley, 1954. Pp. 223-258.

Schaller, G. B. *The mountain gorilla.* Chicago: Univ. Chicago Press, 1963.

Schaller, G. B. The behavior of the mountain gorilla. *In* I. DeVore (ed.), *Primate behavior.* New York: Holt Rinehart and Winston, 1965. Pp. 324-367.

Simonds, P. E. The bonnet macaque in South India. *In* I. DeVore (ed.), *Primate behavior.* New York: Holt, Rinehart and Winston, 1965. Pp. 175-196.

Skinner, B. F. The phylogeny and ontogeny of behavior. *Science,* 1966, **153,** 1205-1213.

Southwick, C. H. An experimental study of intragroup agonistic behavior in rhesus monkeys *(Macaca mulatta). Behaviour,* 1967, 28, 182-209.

Southwick, C. H., Beg, M. A., & Siddiqi, M. R. Rhesus monkeys in North India. *In* I. DeVore (ed.), *Primate behavior.* New York: Holt, Rinehart and Winston, 1965. Pp. 111-159.

Sugiyama, Y. Social organization of hanuman langurs. *In* S. A. Altmann (ed.), *Social communication among primates.* Chicago: Univ. Chicago Press, 1967. Pp. 221-236.

Chapter 3

Vision[1]

Russell L. DeValois

Department of Psychology,
University of California, Berkeley

and

Gerald H. Jacobs

Department of Psychology,
University of California, Santa Barbara

I. INTRODUCTION

It has often been noted that primates in general are extremely "visual" animals. Many mammals (most carnivores, for example) depend very largely on olfactory cues to locate and recognize friend, foe, or food. Others (bats, for example) avoid obstacles and recognize objects largely from auditory signals. Most primates, however, depend almost entirely upon their vision both for

[1] This review and the research reported by the writers have been supported by research grants GB-15969 and GB-12303 from the National Science Foundation and EY-00014 from the National Eye Institute, U.S. Public Health Service.

orientation in space and for recognition of objects. An anatomical counterpart to this visual emphasis is that in primates a very large proportion of the input to the whole central nervous system comes up the optic nerve.

Field studies of primates in their natural habitats have given us important information about the role played by vision. No attempt will be made in this summary of the vision literature to review such field studies. One finding that might be noted, however, is that much communication among primates occurs by the exchange of purely visual, rather than auditory or osmic, information. A common example of this is the submission signal given by one animal to another by drawing back the lips to reveal the teeth; this may take place without any concomitant auditory signal.

There are a number of reasons for studying the vision of monkeys, and the varied types of experiments that have been carried out reflect, to some extent, the diverse goals. One justification for such studies is merely their intrinsic interest. Monkeys and apes are fascinating animals, particularly to their near relatives, man. Numerous studies of primates have been oriented toward discovering as much as possible about the characteristics of some particular species of monkey as an end in itself. In other cases, the study of monkey vision has been undertaken with the hope of understanding the evolution of vision. This would be of particular pertinence to the study of color vision and of binocular vision in different primates. Both of these aspects of vision have evolved, to some extent, within the primates; a comparative study of those existing primates that approximate the different evolutionary levels might tell us something about that evolution. Finally, some workers study primate vision because of the possibility of investigating the underlying physiological mechanisms in nonhuman primates to an extent that is impossible in man. The close parallel between the nonhuman primates and humans would also minimize the difficulties in generalization of the results of such studies to man. A comparative approach is of particular value in this regard, for if different primates can be found which differ in certain respects in their visual behavior, it would be possible to examine how they also differ in the organization of their visual systems. One could thereby come to more firm conclusions about behavioral–physiological relationships than would be possible from the study of one species alone (DeValois & Jacobs, 1968).

II. BRIGHTNESS VISION

Brightness is a basic and ubiquitous property of the visual response. Dozens of experiments on human observers have shown that brightness is determined not only by the physical characteristics of the stimulus itself but also by contextual cues and by the visual history of the organism—recent and distant, and sometimes by future events. Despite this wide range of response determinants, most investigators have studied only the major variables of the wavelength composition and the intensity of the stimulus. Such is particularly true for nearly all brightness experiments with nonhuman primates. The relevant research will, therefore, be grouped into measures of spectral sensitivity and studies of brightness discrimination and contrast.

A. Spectral Sensitivity

Radiant energy that is not absorbed by the retinal photopigments does not contribute to vision. Since the efficiency of absorption by any photopigment varies as a function of the wavelength of light, the sensitivity of the organism to light is necessarily dependent on wavelength as well as intensity. The measurement and meaning of functions relating visual efficiency to wavelength for nonhuman primates is the subject of this section.

1. DEFINITION, UTILITY, AND METHOD

The reader of the vision literature finds several terms that seem to describe the same visual data: luminosity, visibility, radiant luminous efficiency, and spectral sensitivity. Indeed, a recent authoritative text (Graham, 1965) concludes that these terms can be taken as synonymous. "Luminosity" and "sensitivity" are the terms most frequently used in connection with determinations based on discriminative responses from a behaving organism, and thus are deemed most appropriate for use here. The most general definition of a spectral sensitivity (luminosity) function is the reciprocal of energy at different spectral locations required to produce a constant visual effect. It is important to note that such a function does not necessarily describe spectral brightness variation, although it is effectively that in many experimental situations.

There are several ways in which information from spectral luminosity functions has been used. Luminosity functions have frequently provided the basis for inferring other visual capacities, in particular about the presence and nature of color vision. It is, unfortunately, still necessary to point out that the presence or absence of color vision cannot be concluded from luminosity functions *per se*, although such measures may be useful in describing the type of color vision. (Tests of color vision are discussed in detail in Section III.) Luminosity functions may also provide information about visual mechanisms. For instance, many investigators have attempted to estimate the absorption characteristics of the retinal photopigments from luminosity measurements. In these ventures it must be remembered that (a) some of the spectral selectivity seen in the luminosity function is the result of light absorption by other than the retinal photopigments, and (b) the processing in the nervous system that results in the luminosity function may alter considerably the outputs from the photopigments. Finally, luminosity functions are commonly used as a basis for equating light stimuli for relative visual efficiency in studies where the intent is to rule out this relative efficiency as a discriminative cue. This procedure is of obvious importance in the study of color vision.

A variety of procedures have been used with humans to determine spectral luminosity functions and many of these have been adapted for use with nonhuman subjects. Details of these procedures will be presented along with the relevant results in later sections to follow. The point that needs to be made here is that in some cases, the form of the luminosity function is critically dependent on the method used to generate the function. This is particularly so in those cases where the luminosity function is produced by the operation of more than one retinal photopigment; for example, in the usual photopic luminosity function. In this case, different psychophysical techniques tend to promote different combinatorial rules. The reason for emphasizing this method-dependency is to alert the reader to the difficulties involved in comparing luminosity functions obtained with different techniques—particularly in view of the widespread tendency (including that of the authors) to refer to *the* luminosity function of some species.

2. SCOTOPIC SPECTRAL SENSITIVITY

Because most primate visual systems are duplex, it is convenient to consider the scotopic (dim light) and photopic (daylight) luminosity functions separately. Two features of these functions are of interest here: (a) their shapes (spectral locations of peaks, depressions, etc.) and (b) their heights, indicating differences in the absolute sensitivities of the various species investigated.

a. Hominoidea
Apparently no measurements have been made of scotopic sensitivity for any of the apes.

b. Cercopithecoidea (Old World Monkeys)
Macaca. Blough and Schrier (1963) (see also Schrier & Blough, 1963) reported scotopic luminosity functions for three rhesus monkeys *(M. mulatta)* obtained by means of a continuous tracking technique. In this experiment, the dark-adapted subject was trained to press one or the other of two levers so as to maintain a monochromatic test patch at threshold visibility; depression of one lever increased and depression of the other decreased the intensity of the light. In this way, scotopic sensitivity was measured at each of 16 wavelengths between 400 and 700 nm. Two comparisons were presented (Fig. 1): (a) a single human observer run in the same test situation produced a scotopic curve nearly identical to that of the standard Commission Internationale de l' Eclairage (CIE) scotopic luminosity function for humans; (b) the functions for the three

monkey observers were very similar to each other and also to that of the human observer, except at wavelengths longer than 560 nm, where the monkeys showed more sensitivity than did the human observer. Recently, DeValois, Polson, and Morgan (in press) also measured scotopic sensitivity of cynomolgus *(M. fascicularis)* and pigtailed *(M. nemestrina)* macaque monkeys.

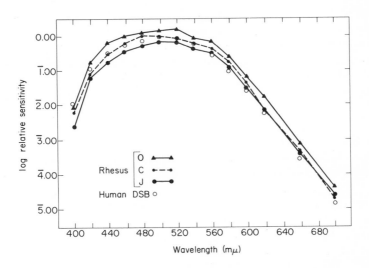

FIG. 1. Scotopic spectral sensitivity curves for three rhesus monkeys and one human tested in the same apparatus. (From Blough & Schrier, 1963.)

Their subjects learned to select a large, slowly flickering light from three nonflickering lights; the energy required for detection was used to determine luminosity. They found very close agreement between the functions for the macaque monkeys, the functions obtained from human observers tested in the same apparatus, and the CIE scotopic function. The difference in sensitivity for the long wavelengths seen by Blough and Schrier does not appear in these latter measurements. The smooth nature of the curves obtained in both of these studies, and their correspondences to the human functions, suggest that macaque scotopic luminosity is based on the same single photopigment as it is for man. There were also no apparent differences in scotopic sensitivity among the three species of macaques examined.

Observations on the absolute scotopic sensitivity of the macaque have been made in three studies. In the DeValois *et al.* experiment, macaque observers were found to be, on the average, about 0.4 log units more sensitive than comparably-tested human observers. On the other hand, Monjan (1966), using an increment-threshold test, and Blough and Schrier (1963) (see also Schrier & Blough, 1963), in the study mentioned earlier, found macaque and human observers to have approximately the same scotopic sensitivity. There is no way to reconcile these variant results, but it should be remembered that there may very well be differences in response criteria that man and monkey adopt in these different tests, and that the details of the testing situations must differ for the two types of subjects.

Cercocebus. Adams and Jones (1967) determined scotopic luminosity functions for four mangabeys using the flicker tehcnique described above. The functions so determined were found to be very close in form to those obtained from human observers tested in the same situation (with peaks at about 510 nm). Both species in this instance were less sensitive to the short wavelengths than were the human observers upon whom the standard CIE function is based.

c. Ceboidea (New World Monkeys)

Cebus. Morgan and DeValois (in preparation) have made measurements of scotopic sensitivity of cebus monkeys, using the flicker technique. The resulting curve closely approximates the human curve.

Saimiri. Scotopic luminosity has also been measured with flicker techniques in the squirrel monkey (Jacobs, 1963). Once again, there is close correspondence between the form of the scotopic function for the squirrel monkey and the CIE function, except below about 500 nm, where the curve for the squirrel monkey is somewhat depressed relative to the CIE.

d. Prosimii

Galago. The galago, or bush baby, has an all-rod retina containing only a single photopigment (Dartnall *et al.*, 1965) and hence luminosity measurements for this animal can be justifiably included here, even though no specific effort was made to eliminate photopic vision. Silver (1966) determined the spectral sensitivity for one of these primates in a two-choice discrimination problem where the animal initiated each trial by making an observing response. The luminosity function resulting from this procedure was closely similar in form to that obtained from a human observer. The slight displacement of the peak of the galago luminosity function from the peak of the absorption function of its retinal photopigment was attributed to light reflected from the retinal tapetum.

e. Conclusion

Scotopic luminosity functions from the primate species so far investigated are all quite similar in form, reflecting the similarity in the underlying retinal photopigment. Whether or not there are systematic species differences in absolute scotopic sensitivity is an interesting question, which, for the moment, remains unanswered.

3. Photopic Spectral Sensitivity

In determining photopic sensitivity some means must be employed to eliminate any scotopic contribution. The techniques utilized with monkey subjects include many of the methods previously demonstrated to be effective for human observers: providing a high level of ambient illumination, using a test stimulus that is added to a moderately intense background light, requiring the animal to discriminate a rapidly flickering light, and making sensitivity measurements along the cone limb of the dark-adaptation function. Unlike the scotopic luminosity functions discussed above, photopic functions are frequently complex in form and within-species differences are very likely to be methodologic in origin.

a. Hominoidea

Although it is frequently assumed that the photopic sensitivity of apes should be similar to that of man, there is only very meager experimental evidence. Grether (1940) measured the spectral limits for photopic sensitivity for chimpanzees *(Pan)* and found these values to be similar to those for humans.

b. Cercopithecoidea

Macaca. As usual, the macaques have been the center of experimental attention. In his pioneering study of monkey vision, Grether (1939) measured the photopic sensitivity of one rhesus monkey. The animal was required to choose a lighted box in preference to a dark one, the intensity and wavelength of the positive stimulus being variable. Photopic adaptation was achieved by incandescent illumination of the test chamber. The functions resulting from this procedure were quite similar for the rhesus and three human observers: all showed a sensitivity peak around 550 nm and a large dip in sensitivity centered at 580 nm. Grether suggested that the latter feature probably resulted from the presence of the yellowish adaptation light.

Schrier and Blough (1966) utilized the tracking procedure described above to derive photopic luminosity functions for a rhesus and a stumptailed macaque *(M. arctoides).* In this case, their subjects generated sensitivity functions during the first 3 min of dark adaptation, i.e., along the cone limb of the function. Photopic sensitivities were found to be about the same for the two species, both showing peak sensitivity at about 520 nm. In comparison with the CIE photopic function for humans, the two monkeys were considerably more sensitive to the short wavelengths and slightly more sensitive to the very long wavelengths.

Sidley and co-workers (Sidley, Sperling, Bedarf, & Hiss, 1965; Sidley & Sperling, 1967) measured the photopic luminosity function of rhesus monkeys in an increment-threshold situation, using two different behavioral tasks. In one, the animal learned a shock avoidance response, where the response was cued by the presentation of the discriminative stimulus; in the other task, the monkey was trained to release a lever within a brief interval following stimulus presentation in order to receive appetitive reinforcement. Functions generated in the increment-threshold context, for both man and monkey, show multiple sensitivity peaks. For both rhesus and man, the main sensitivity peak occurs at 530 nm with subsidiary peaks at around 600 and 440 nm. Similarly, for both species, sensitivity depressions appear at 480 and 580 nm. Statistically, Sidley and his colleagues found no differences in photopic sensitivity between rhesus and human subjects. Their experiments are noteworthy for improved spatial and temporal control of the stimulus.

Recently, DeValois and co-workers (DeValois, 1965; DeValois & Jacobs, 1968; DeValois *et al.,* in press) determined the photopic luminosity functions for cynomolgus and pigtailed macaques by measuring the intensity of various spectral lights required to produce fusion of a rapidly flickering stimulus. This method is predicated on two facts: (a) at high flicker rates, considerable light is required to stay above fusion threshold, thus maintaining photopic adaptation, and (b) that the scotopic system cannot follow rapidly flickering light. Peak sensitivity for the monkey subjects was found to be in the 520–560-nm region. In comparison with human observers run in the same apparatus, the photopic sensitivity of the macaques was enhanced in the short wavelengths and somewhat depressed at test wavelengths longer than 600 nm.

The above statements emphasize the variability of the photopic luminosity functions for the macaque monkey. This variability should not mask the central result of all of these studies: When macaque and human luminosity functions are determined by the same techniques and with the same apparatus, the differences between the functions are small and, probably, not significant. Evidence for this conclusion is given in Fig. 2, which shows comparable luminosity functions for man and macaque determined in two different studies. As a subsidiary point there is, so far, no evidence for differences in photopic sensitivity among the four macaque species that have been investigated.

The variability in photopic luminosity in different test situations might be interpreted as follows. Previous evidence (DeValois, Abramov, & Jacobs, 1966) suggests that there are six different types of neural paths from the visual receptors to the brain in the macaque monkey. In one pair of pathways (nonopponent excitators and inhibitors), the outputs of different cone types are summed to provide black–white information. In the remaining pathways (spectrally opponent), the outputs of different cone types are subtracted from each other to provide chromatic information. Sensitivity presumably reflects the overall activity of all cell types, but certain stimulus situations emphasize the contribution of one or the other of the types of neural pathways. In the flicker situation, the contribution of the opponent cells would be minimal, since the chromatic systems cannot follow a rapidly flickering light; under these circumstances the photopic luminosity curve matches the output from the nonopponent cells (DeValois *et al.,* 1966). In the increment-threshold situation,

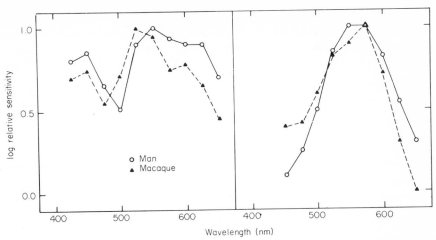

FIG. 2. Photopic luminosity curve for macaques and man. On left: from Sidley *et al.* (1965), based on an increment threshold technique. On right: from DeValois *et al.* (in press), based on a flicker technique. Note that although these two techniques produce widely different luminosity functions, the macaque and human subjects are very similar in each experiment.

on the other hand, the steady white background light would tend to adapt out the nonopponent cells so that the photopic luminosity under these circumstances more closely matches the summed output of the spectrally opponent cells (note the resemblance of Fig. 3 in Sidley *et al.*, 1965, to Fig. 18 in DeValois *et al.*, 1966).

While the studies above cited agree in showing man and macaque to have the same relative photopic luminosity function, they disagree completely on the absolute sensitivities of these species. Rhesus monkeys have been found to have photopic sensitivities that are substantially lower than those for man (Grether, 1939) or to not differ significantly (Sidley & Sperling, 1967), whereas the pigtailed macaque has been found to have considerably (Monjan, 1966), or only slightly (DeValois *et al.,* in press) higher photopic thresholds than those for human observers.

Finally, mention should be made of a recent study (Sperling, Sidley, Dockens, & Jolliffe, 1968) in which photopic luminosity functions were obtained from rhesus monkeys under conditions of neutral and chromatic adaptation. The point of such studies is to provide a means of separating the component processes that combine to produce the photopic luminosity function. The results of the initial attempts of this type suggest that the principles of combination are complex, and may be different for different spectral regions.

Cercocebus. Photopic luminosity functions for four mangabeys were obtained using the flicker technique (Adams & Jones, 1967). The peak sensitivity was found to extend from about 500 to 580 nm. In comparison to human controls, the mangabeys showed enhanced sensitivity to the short wavelengths and a 0.3 to 0.4-log-unit depression in the long wavelengths.

c. Ceboidea

Cebus. In a study referred to previously, Grether (1939) reported photopic luminosity functions for two cebus monkeys. The derived curves are difficult to characterize—the most notable feature being a large depression in sensitivity centered at 540 nm—and Grether himself dismisses the results as being of questionable value. On an absolute scale, the cebus subjects were found to be as much as 2 log units less sensitive than human controls. This is far out of line with other results. More recently, measurements of cebus photopic sensitivity have been made using the flicker technique (DeValois, 1965; Morgan & DeValois, in preparation). It was found that the form of the sensitivity curve for this monkey agrees quite closely with that of macaque and human observers tested similarly with, at most, a slight relative loss beyond 600 nm for the monkey.

Saimiri. Photopic luminosity functions for five squirrel monkeys, obtained from flicker measurements, all showed sensitivity peaks at spectral locations close to those seen for man (Jacobs, 1963). However, relative to the same base, these photopic functions showed an elevation in sensitivity to the short wavelengths and a substantial loss in sensitivity (average of 0.4–0.7 log units) to wavelengths beyond about 600 nm.

d. Prosimii

The tree shrew *(Tupaia)* apparently possesses an all-cone retina. Behavioral measurements by Polson (1968) show no evidence for a Purkinje shift in the common tree shrew *(T. glis)* and thus verify the lack of any rod contribution to the vision of this animal. Relative sensitivity, measured by the flicker technique, was very little different from that of human observers. This result is of particular interest in view of the nature of color vision in this species (discussed in Section III). On the other hand, the same experiment shows the tree shrew to have substantially and uniformly more overall sensitivity than man. The difficulties of interpreting differences in absolute sensitivity were alluded to above and apply with equal force here.

B. Brightness Discrimination and Contrast

1. BRIGHTNESS DISCRIMINATION

The term brightness discrimination might be applied to any experimental situation in which the luminance magnitude required to produce some criterion response is determined. Many of the studies of spectral sensitivity mentioned above could, therefore, be considered under this heading. Classically, however, investigators of brightness discrimination have avoided the manipulation of stimulus wavelength and have concentrated on the manner in which detection is affected by the duration and size of the test stimulus, and by various features of the adaptation state of the subject. Experiments of this sort on nonhuman primates have been exceedingly rare.

Typical situations for studying brightness discrimination are a bipartite field with luminances independently variable in the two halves of the field, or a single test field that receives a change in luminance for a brief period of time. Both of these general situations have been employed with monkey subjects. In an early study, Crawford (1935) trained rhesus monkeys to select the brighter of two adjacent stimulus windows and determined, at four standard luminance levels, the magnitude of the luminance that had to be added to the standard in order for the animal to detect it reliably. Brooks (1966) used the second procedure in a study in which three squirrel monkeys were trained to respond to the brighter of two stimulus targets: Both targets were illuminated for a 25-sec period, but during the last second, one of the targets received an increment in luminance.

Discrimination capacities in these experiments were expressed as Weber fractions (the increment in luminance necessary for criterion discrimination ΔI, over the background luminance I). This ratio $\Delta I/I$ is then plotted as a function of the background luminance. Many experiments on human observers have demonstrated that the size of the Weber fraction declines from a relatively high value at dim background levels to a more-or-less constant value over a range of moderate luminances. Both rhesus and squirrel monkeys were found to produce this same general function. In addition Crawford directly compared rhesus and human observers and found very similar discrimination performances in the range from 0.78 to 55.3 ml, but distinctly better performance for man at a very low background level (0.086 mL). In view of all the other evidence for similarities in macaque and human vision, this latter difference may well be a matter of performance rather than capacity.

2. Discriminations Based on Luminous Flux

The general paradigm for brightness discrimination requires that the animal discriminate between two (or more) like targets that differ in luminance. In such a situation, the targets to be discriminated actually differ in both luminance (light per unit area) and luminous flux (total amount of light); each of these features is potentially useful as a basis for discrimination. That luminous flux may, in fact, be utilized as a cue for discrimination was suggested some time ago by Klüver (1941), who reported that monkeys with damage to the occipital lobes could no longer make visual discriminations on the basis of luminance differences, but retained the capacity to discriminate luminous flux.[2]

Schilder, Pasik, and Pasik (1967) have reported an experiment which implies that, at least under certain circumstances, monkeys in a typical brightness discrimination problem may utilize differences in luminous flux rather than differences in luminance as the cue for discrimination. The investigators first trained rhesus monkeys on a problem in which the two stimuli to be discriminated differed in both luminance and luminous flux. After mastery of this problem, the animals were retrained on a discrimination problem in which the two stimuli were equal in luminous flux, but differed in both area and luminance. The number of acquisition trials for these two problems did not differ significantly. This result, buttressed by a variety of control conditions, suggested that the difference in luminous flux was the basis on which the monkeys formed their initial discriminations.

It seems clear that both luminance and luminous flux constitute stimulus dimensions that monkeys are able to utilize for discrimination, at least under some circumstances. This fact should be kept in mind when analyzing experiments in which the cue values of luminance and luminous flux differ (for example, where the areas of the discriminative stimuli differ drastically). It might also be noted that, although the extreme conditions under which only luminous flux is a factor (small stimuli, very brief exposures) have been extensively investigated on human observers, the role of flux as a cue for discrimination performance is virtually unmentioned in the literature on human vision.

3. Brightness Contrast and Constancy

A brightness contrast situation is one in which spatial regions having different luminance values are located in proximity to one another, or in which regions possessing different luminance values are juxtaposed temporally. The effects of such placements are to alter the brightness relationships in the field. Although it has been extensively studied with human observers, we can report here only a single study on nonhuman primates concerned with brightness contrast.

Grether (1942) initially trained two chimpanzees to select the brighter of two small squares when both appeared against identical gray backgrounds. Next, he presented two identical gray squares on backgrounds having different reflectances. He assumed that if contrast were present, the stimulus having the darker background would appear brighter and thus would be chosen by the chimp. The results were somewhat equivocal; whereas one animal responded positively to the stimulus on the darker background (contrast prediction), the

---

[2] That this result necessarily implies a different neural basis for flux and brightness discrimination has been challenged (e.g., Gross & Weiskrantz, 1959).

other initially responded in the opposite direction, although with repeated trials both animals eventually behaved in accord with the contrast prediction. In addition, Grether also measured the magnitude of the contrast effect by determining how much light had to be added to a center spot with a bright background so that the animals would choose it over another spot having the same standard luminance but a much dimmer background. The results suggested tha the magnitude of the contrast effect was roughly the same for the chimpanzees and several human subjects.

The difficulty of investigating brightness contrast in nonhuman subjects is clearly exemplified by Grether's study. With humans it is possible, through instruction, to be fairly sure that the subject's response is based on the appearance of the test stimulus. On the other hand, with animals well trained to select, say, the brighter of two stimuli, the sudden appearance of a large brightness difference in the background stimulus is as potent a cue for the choice as is the stimulus that the experimenter is designating as "test." Studies of contrast with monkey subjects will require different experimental designs than those for human subjects. In view of the current considerable interest in contrast mechanisms, the development and test of such designs would be most useful.

In the face of the array of stimulus characteristics that affect brightness, the brightness relations among objects nevertheless tend to remain relatively constant. There are experimental indications that such constancy relationships are also present among the nonhuman primates. For example, Köhler (1915) trained a chimpanzee to choose a white paper rather than a black one, and found that the animal continued to select the white one even when the black was illuminated so as to reflect many times as much of the light as did the white. Likewise, Klüver (1933) showed that cynomolgus macaques trained to select the brighter or the darker of a pair of cards continued to respond relatively even when the brightness of the two stimuli was so varied that the absolute luminance of the light one was the same as the darker one had previously been. The limits of and conditions for brightness constancy have not received much investigation with monkeys, but it is unlikely to prove any less complicated than for human subjects.

III. COLOR VISION

A. Methods of Study

There are few areas in science in which the pitfalls exceed those to be found in the study of color vision. Since experimenters only gradually discovered what controls were required before one could specify whether or not an animal had color vision, much of the early literature is quite uninformative. Even today there is widespread misinformation about color vision, and a frequent confusion of color vision with spectral sensitivity, for instance.

Light from each point in space can vary in only two respects: in intensity (or number of photons), and in wavelength (or energy level of the photons). Both of these features are potentially useful to the organism in detecting the presence of an object at that point in space.

Every eye that has evolved has probably developed the capability of extracting intenstiy information from the stimulus, but only a few can extract the information necessary to make wavelength discriminations. The initial requirement for detecting the intensity of light is a receptor containing a number of photopigment molecules that can operate sufficiently independently of each other to generate different amounts of neural activity depending upon how many molecules capture light. Extracting information about wavelength is a much more difficult task, because a photopigment has no information about the wavelength of the light that it has absorbed. While photons of some wavelengths are more likely to be absorbed than are others, once absorbed, every photon has exactly the same effect. A single receptor containing many molecules of a single photopigment type can thus signal the intensity of the light, but cannot provide the organism with any information about the wavelength of the light which it has absorbed. To extract wavelength information, an organism requires at least two receptors containing photopigments that have different spectral sensitivities. A red light would thereby be potentially distinguishable from a green light, regardless of their relative intensities, because the long-wavelength-absorbing pigment would always respond more to the red light than would the short-wavelength-absorbing pigment; and the latter pigment would always respond to the green light more than would the former. The crucial information which the nervous system should obtain from the receptors to specify the wavelength of the light is thus not the amount of activity in any one receptor, but the relative activity rates of receptors containing different photopigments. Those animals having color vision have exactly this sort of neural organization (DeValois & Jacobs, 1968).

One of the factors which has led students of animal vision into uninformative experiments and fallacious conclusions has been a confusion between differential spectral sensitivity and color vision. Virtually any pigment, including all the photosensitive pigments found in plants and animals, absorbs light differentially, depending on wavelength. The rod pigment, rhodopsin, for instance, is maximally sensitive to about 500 nm, with the sensitivity falling off to about one tenth of that at 600 nm. Two lights of 500 and 600 nm, respectively, equated for physical intensity will be quite differentially absorbed by this pigment. An animal possessing only rhodopsin could quite readily tell these lights apart: They would differ in brightness. Since many animals can detect very small brightness differences, almost any combination of 500- and 600-nm lights could be distinguished by an animal whether it had color vision or not. An animal could only be said to have color vision, however, if it were capable of differentiating these two lights when their relative intensities were adjusted to equal brightness for the animal.

The necessity of ensuring that an animal in a color discrimination test is not choosing on the basis of brightness differences has been recognized, at least since the study of Kinnaman (1902). While most experimenters have been aware of the problem, the methods used for controlling the luminance of the lights have not always been ideal. Some early experimenters merely made the lights equally bright to their eye. This would be perfectly adequate if the luminosity function of the animal were the same as that of the experimenter, but if the animal has a different type of color vision it might very well also differ in its sensitivity to different wavelengths. Furthermore, if the animal had no color vision, it might well notice smaller differences in brightness between two different colored lights than would the experimenter. Our great sensitivity to color differences makes it difficult for us to distinguish brightness differences between two lights that are of different colors. A color-blind animal would not have this difficulty.

A common technique for brightness control has been to introduce a large range of brightness differences. Thus, in an achromatic–monochromatic discrimination, an achromatic array ranging from white through the grays to black might be used. Although such a range would be quite unnecessary if prior brightness measurements had been made, it would be an adequate brightness control *provided* that careful records of errors were maintained. However, 80% of the achromatic stimuli, for instance, might be brighter for the animal than the colored one so that the animal could achieve 80% correct discrimination, simply by choosing the dimmer light, without any use of color vision. Clearly, in this case, one would have to know which of the stimuli were missed. Such a shotgun approach to the control of brightness would be particularly unfortunate in studies of the nature of color vision in an animal (e.g., saturation discrimination, or hue discrimination) which require more precise procedures. In these experiments, for instance, a wide intensity variation would produce major artifacts, since hue varies with light intensity to some extent (Bezold–Brücke effect). The method of choice, therefore, would be to measure for each animal the relative brightness of the chromatic and achromatic stimuli to be used and to use these values (plus a sufficient intensity variation to compensate for possible experimental errors) to equate the various lights.

1. Tests for the Presence of Color Vision

The most fundamental question about the color vision of certain animals is, of course, whether or not it is present. Unfortunately, it is true that studies of the color vision of most primate species have not gone beyond the mere question of its presence, although it is of equal interest to determine the nature and degree of the color vision of an animal which possesses this capability. An animal may have color vision, but of such limited extent that it could not play an important role in its visual behavior.

An individual who has no color vision is a "monochromat": he can match any comparison light by varying just *one* dimension of *one* test light, such as its intensity. A "dichromat" requires a mixture of *two* test lights, each being variable in one dimension, to match any given comparison stimulus, and a "trichromat" *three*. Only a monochromat is, of course, completely color-blind. A test for the presence of color vision is, in this sense, a test for possible monochromacy.

Almost any color vision experiment—hue discrimination, saturation discrimination, color-mixing, color contrast, etc.—might serve as a means of identifying a monochromat. The two most widely used tests are really extreme cases of hue and of saturation discrimination.[3] In the former, the animal's ability to discriminate between two widely separated wavelengths is examined. A monochromat can match one light to another by varying its intensity. In a discrimination experiment, he would thus not be able to discriminate between them when they were equated for brightness. However, a single failure on the part of an animal to solve such a problem would not be definitive evidence for absence of color vision, because a dichromatic individual is also incapable of discriminating between pairs of widely separated wavelengths in one spectral range. The most common human dichromats, deuteranopes and protanopes, cannot discriminate among wavelengths in the long-wavelength part of the spectrum. Thus a failure to discriminate between green and red (to say nothing of green and yellow, or yellow and red) would not necessarily indicate monochromacy. The third main variety of human dichromats, tritanopes, confuse the blues and greens, and, therefore, a failure of wavelength discrimination in this spectral region would also not necessarily indicate monochromacy. The definitive test would be one that showed that an animal could not discriminate between green and *either* of the spectral extremes: No animal whose vision covers the same spectral range as ours and who has color vision would fail on such a test.

The other common test for color vision is a white versus monochromatic discrimination (the extreme case of a purity discrimination test). For mono-

---

[3] We follow the usual visual terminology in referring to these as "hue" and "saturation" discrimination tests, although hue and saturation are response, not stimulus terms. The animal or human is actually being asked to discriminate between lights of different wavelength or purity; the tests should therefore be termed wavelength or purity discrimination tests, respectively. This distinction is particularly relevant with regard to wavelength discrimination by a dichromat, because much of his discrimination of wavelengths is actually based on saturation, rather than hue differences.

chromats the whole spectrum is "neutral," i.e., indiscriminable from the white–gray–black dimension, given an adequate brightness match. The difficulty with establishing monochromacy with a spectral-white discrimination is that dichromatic individuals also have a "neutral point" in the spectrum: There is a narrow band of wavelengths that stimulates the two photopigments in exactly the same proportion as does white light, and these wavelengths are indistinguishable from white light. The neutral point for human deuteranopes and protanopes is the blue-green at about 490–500 nm, depending on which of many possible stimuli are chosen for the "white" standard. The higher the color temperature of the white, the lower the neutral point. The neutral point for protanopes is at a somewhat lower wavelength than for deuteranopes, but the difference is too slight to serve as a reliable means of distinguishing between them. Human tritanopes have a neutral point in the yellow wavelength region at 570 nm (with a second neutral point likely in the deep blue). A choice of one of these spectral bands in a white-monochromatic test for color vision might thus lead to a confusion between monochromats and dichromats. No human dichromat has a neutral point in the red, nor would any animal with anything approximating the human photopigments; thus the ability to discriminate between white and red would be a reasonable test for monochromacy, although at least two colors should be tested for convincing proof of a lack of color vision.

2. Tests for Different Types of Color Vision

We know from extensive studies on humans that individuals who have color vision show numerous differences in the type. Table I gives the traditional classification of types of color vision. Following an ancient, if not honorable, tradition in vision, we will ignore the 50% or more of human color-defectives

TABLE I

Varieties of Color Vision

Number of variables	Subclasses
Trichromats	i. normal
	ii. protanomalous
	iii. deuteranomalous
	iv. tritanomalous
Dichromats	i. protanopes
	ii. deuteranopes
	iii. tritanopes
Monochromats	i. typical (rod)
	ii. atypical (mono-cone)

who cannot be fitted easily into this classification scheme (many of these have color defects acquired as a result of various diseases).

The most definitive test to distinguish dichromats from trichromats is the neutral-point test: Trichromats can discriminate any monochromatic light from white, whereas, as was mentioned in Section III, A, dichromats have a neutral point. Since such a systematic search across the spectrum of the ability to discriminate spectral light from white can also be a convenient way of discovering the presence of color vision, it can be recommended as the initial test to be carried out on an animal whose color vision is being examined for the first time.

If a neutral point were found, its spectral location would enable one to distinguish a tritanope from a protanope or deuteranope, but would not permit an unequivocal differentiation between the latter two types of dichromats. For this differentiation, the best evidence is probably not from a test of color vision at all, but from the photopic spectral sensitivity functions: Protanopes are very much less sensitive to long wavelengths than are deuteranopes (or normal trichromats). Another, more complicated, test is that of the isochromatic confusion lines. It has been shown (Pitt, 1944) that, whereas protanopes confuse purple lights with blue, deuteranopes confuse these same colors with blue-green and green. Thus a test of the ability to discriminate between a purple light and various parts of the spectrum provides a powerful way to distinguish these types of defect in color vision.

Human trichromats differ among themselves in their color vision to a considerable extent, with those sufficiently far from the mean being classified as anomalous. A very useful means of distinguishing between the two most common varieties of anomalous trichromats is the Nagel Anomaloscope. In this device, which can be easily adapted for the study of animals (Trendelenburg & Schmidt, 1930), the particular mixture of a red and a green which matches (i.e., is indiscriminable from) a yellow is found. Protanomalous individuals require more red and deuteranomalous more green in the mixture than do normals.

A similar color mixing arrangement might be devised in the blue-green range to distinguish tritanomalous from normal. However, here one would find that normals show no true match point (point of complete indiscriminability) since a blue-green mixture is desaturated relative to monochromatic light of inter-mediate wavelengths. Nonetheless, one should observe some drop in performance at some intermediate point.

3. Tests for Amount of Color Vision

Some discussions of color vision in different animals seem to assume that the possession of color vision is an all-or-none thing. The long drawnout attempt to find some evidence for color vision in the cat was, to a large extent, based on the desire to use physiological data from the cat's visual system to understand human vision. However, even if the cat does have some color vision (and there is little doubt that it does) it is clearly so little *quantitatively* as to be completely unlike the human visual system. Thus a cat must be trained for hundreds or even thousands of trials before it gives any evidence of being able to discriminate between a light which is 100% white and one which is 100% monochromatic blue; a macaque monkey, on the other hand, can quite readily distinguish between a light that is 100% white and one that is 99% white and 1% blue. This is quite a different question from that of the *type* of color vision, as discussed above: The cat may be trichromatic (the data are not clear on this point) just as is the macaque, but there is a *huge* quantitative difference between the cat's color vision and that of many of the primates.

One of the best measures of the degree of color vision is the saturation of the spectrum. The smallest admixture of monochromatic light in a white light that is discriminably different from a pure white light for the animal can be determined. An animal with excellent color vision, such as man, can distinguish the presence of extremely small amounts of monochromatic light, particularly at the spectral extremes. Such a test conducted at different wavelengths across the spectrum characteristically shows that the relative saturation of the spectrum varies considerably; for a normal human trichromat the region around 570 nm is much less saturated than other parts of the spectrum. It is of considerable interest to know the form of the saturation discrimination curve for other primates. Such a test also provides evidence on the type of color vision, since the least saturated part of the spectrum is not at 570 nm for most anomalous trichromats and dichromats (the neutral point of a dichromat is, of course, a region of zero saturation).

A wavelength discrimination test also provides evidence for the degree of color vision. An individual with excellent color vision can distinguish a very small difference in wavelength whereas an individual with weak color vision requires a very much larger wavelength difference for the same level of discrimination. To compare the cat and the macaque again, one can note that the cat has difficulty discriminating the 150-nm difference in wavelength between 600 and 450 nm, whereas a macaque can discriminate the 2-nm wavelength difference between 600 and 598 nm. For any animal, the ability to discriminate wavelength differences varies across the spectrum; the shape of the wavelength discrimination curve tells one much of interest both about the amount and the nature of an animal's color vision.

4. SUMMARY OF METHODS

The necessity to control the intensity variable in measuring color vision was emphasized: It is by far preferable to measure the photopic luminosity of the spectrum and the equivalent white and to use these values to equate different lights for equal brightness. The presence of color vision in an animal can be determined from various tests. Determinants of whether or not various monochromatic lights can be discriminated from white allow a decision as to the presence of color vision, and whether the color vision is trichromatic, dichromatic, or monochromatic. Anomalous trichromats can be distinguished from normals by means of color-mixing experiments. It was emphasized that color vision varies not only qualitatively in different individuals, and perhaps species, but also quantitatively. Studies of saturation and hue discrimination can be used to assess the degree of color vision possessed by an animal, and also provide much other information of interest about its visual behavior.

B. Survey of Work on Color Vision

1. HOMINOIDEA

a. Pan. Although only the chimpanzee has been extensively studied, there is little doubt that all of the apes have color vision. Although Kohts (1928) did not directly test for color vision in the chimpanzee that she studied for 2 years, the behavior of the animal in certain visual tests was such as to indicate not only the presence of color vision, but also the great importance of color in the animal's visual behavior. She used numerous tests based upon a matching-to-sample technique, in which the animal had to choose that one of several stimuli that matched the sample with which it was presented. She reported that from eight spectral stimuli the chimpanzee could readily select the one that matched the color of the model. Although there were no controls to ensure that the animal was not choosing on the basis of brightness differences, the number of stimuli involved, plus the fact that the animal reportedly solved this problem much more easily than one in which the choice had to be based on brightness makes it very likely that color formed the basis of the chimpanzee's choices. Furthermore, Kohts reported that she once gave the animal a collection of 35 black, white, and different colored chips, and soon noticed the chimpanzee spontaneously sorting them by color! The great importance of color in the chimpanzee's visual world was indicated not only by its making color choices much more readily than brightness ones, but also from the fact that when color and form were combined in a sorting experiment, the animal ignored form and sorted the chips by color alone.

The definitive studies of color vision in the chimpanzee were by Grether (1940a, b, c, d, 1941a, 1942). These excellently designed experiments showed not only that the chimpanzee has color vision but that its color vision closely parallels both in quality and extent that of normal human trichromats.

In these experiments, Grether used a two-choice discrimination apparatus in which the food cups were covered with white discs onto which the stimulus lights were projected. Beams from either a white-light source or a mono-chromator could be employed. The animal displaced the appropriate card to get food.

Hue discrimination was tested at three spectral points for four chimpanzees and four humans (Grether, 1940a), and again at some 20 points across the spectrum with three chimps and one human (Grether, 1940d). The hue discrimination function of the chimps closely resembled that of the normal human observers tested in the same apparatus. Both showed the familiar double minimum: at about 600 and at 470 nm (although virtually all other studies of normal human observers have found the short-wavelength minimum to be at 490–500 nm; the reason for this discrepancy is not clear). Discrimination of wavelength differences was poorer in the middle of the spectrum and at either spectral extreme. Furthermore, the chimpanzees showed about the same absolute level of discrimination as the normal human observers, except in the deep red region where the difference thresholds for the chimps were somewhat higher. Grether emphasized this discrepancy in the red when discussing his results (arguing that the chimpanzee is intermediate between the cebus monkey, which he found to have a very large difference threshold there, and man), but clearly the more striking finding is the very close agreement between the chimpanzee and man over most of the spectrum.

With the same apparatus and subjects, Grether (1940b) compared chimps and normal humans on two color-mixing problems. One was the Rayleigh match of a red (640 nm) plus green light (560 nm) to equal a yellow. In this test, the beams from the two monochromators now fell on the same food cup, and a white-light beam was employed to illuminate the other food cup. Initially the animals were trained to pick out the red from the yellow light; then increasing amounts of the green light were added to the red until the discrimination dropped toward chance. Then the problem was reversed, the animal now being trained to choose the yellow light when it was paired with the green. Then increasing amounts of red light were added to the green until the performance level again dropped. The cross point of the two functions determined in this way was taken as the red–green mixture that most resembled the yellow. This turned out to be the same mixture as was required for the normal humans. Thus, one can conclude that not only are chimpanzees trichromats, but that they are normal trichromats ("normal" of course being ethnocentrically defined as like most human observers).

The other color-mixture experiment that Grether (1940b) conducted was designed to determine what proportion of orange (610 nm) to blue-green (495 nm) light in a mixture matched white light. The procedure used was identical to the anomaloscope experiment described above, except that unfiltered white light was used instead of yellow and, of course, the monochromators were set to different wavelengths. This experiment showed that the chimpanzees required a slightly greater amount of orange (or less blue-green) in the mixture of orange and blue-green to give white than did the human observers. Grether concludes from this that the chimpanzee has color vision that is very similar to normal man except for a slight loss in saturation at the long wavelengths. One would, however, expect that if a long-wavelength deficiency were present, it would also have appeared in the Rayleigh match. A more attractive explanation of the different complementary mixture for the chimpanzees might be that they have a slightly higher sensitivity to blue, rather than a loss in the red. This could result from neural differences or, more likely, from differences in preretinal absorption. In man, the lens yellows with aging, progressively cutting down transmission of short wavelengths. The chimpanzees might have naturally had yellow lenses or, being adolescent or preadolescent animals, might be expected to have less lens pigmentation and, therefore, a higher sensitivity to blue relative to the human observers with whom they were compared.

In still another study of chimpanzee color vision, Grether (1941a) examined the saturation of different parts of the spectrum. Two white lights (of $4000° K$) illuminated the two food-boxes. The amount of monochromatic light at each of 17 wavelengths that had to be added to one of the whites for detection of the color was determined. The resulting curve showed that the spectral extremes were very saturated for both chimpanzees and normal human observers (a very small amount of monochromatic light had to be added to white light to be detected). The yellows, however, were in both cases very much less saturated, the minimum saturation being at 575 nm for the human observers and at 570 nm for the chimpanzees. Here again one is struck by the very close agreement between these two species, although Grether emphasized the slight discrepancy. In this saturation-discrimination experiment, the absolute saturation as well as the relative saturation of the spectrum, was found to be nearly the same for chimpanzees and man. This would indicate that the degree of color vision, as well as the type, is very much the same in these two species.

In a final examination of the chimpanzee visual system, Grether (1942) tested for the presence and magnitude of simultaneous brightness and color contrast, again testing both chimpanzees and normal human observers. The apparatus had large cardboard sheets, either white or colored depending on the stage of the experiment, surrounding each of two windows that were transilluminated from an optical system. The center and background of each window were covered with tracing paper to give a more uniform surface. In the initial training for the color-contrast experiment, the backgrounds were white, and a faint amount of red light was projected to one window and a faint amount of green light to the other. The animal had to choose the red window. Then test trials were given (with both windows baited) in which the windows were illuminated with white light, but the background was red for one window and green for the other. The performance of all three chimpanzees (and also the human observers) gave evidence of contrast effects; that is, they chose the window with the green background (which induced a reddish appearance on the window). Grether then went on to quantify the contrast effect by adding small amounts of green light to the window with the green background in order to counteract the induced red. As more and more green light was added, the choice dropped progressively from 100 to 0%, which allowed him to measure how much "real" green light just cancels the "induced" red light. It was a considerable amount, and was the same for the chimpanzees and man.

b. Other species.

The only other study of color vision in apes was by Tigges (1963b), who tested one orangutan *(Pongo)* and three gibbons *(Hylobates lar)*. The technique used was that of discriminating colored paper from gray. Four Oswald color papers were used along with 62 gray papers of varying reflectances. In the principal series of experiments, one colored paper was presented with four different grays. If this experiment had been carried out with the grays presented in random order over a number of trials until all had been paired with the color repeatedly, and with the animal having to choose the colored paper to get the reward, the results would have been straightforward and easily interpretable. The actual experiment was much more complicated and almost obscures the result, which is that both of these species have color vision.

The animals were initially trained to discriminate between a particular color and the four darkest grays, with only the color being rewarded. Then they were "tested" with the color against four lighter grays, with any choice now being reinforced. If the subjects chose the colored paper fairly consistently on this "test," the paper was then presented paired with three still lighter grays, etc., through the series. The reason for this combination of discrimination and generalization trials is not presented. In any case, the orangutan performed without any difficulty and clearly gave evidence of color vision. The gibbons had no difficulty discriminating between the colored and the gray papers when they had to do so to be rewarded, but often strayed into choosing the grays on those trials when it made no difference what choice they made. This latter fact, of course, says nothing about their color vision.

It is unfortunate that, except for Grether's extensive and thorough studies of the chimpanzee, we have virtually no information about the nature of color vision in the apes.

2. CERCOPITHECOIDEA

a. Macaca.

Almost all of the studies of color vision in Old World monkeys have been on one genus, *Macaca*. All studies have indicated the presence of excellent color vision in these monkeys.

Kinnaman (1902) was the first to study color vision in nonhuman primates, and was also the first to recognize and state clearly the necessity of controlling brightness in tests of color discrimination, although the controls he used were less than ideal. He first tested the ability of two young macaques to choose one of six glasses covered with red, yellow, green, blue, light gray, and dark gray paper. In successive series of tests, they were trained to choose one after another of these from the group, which they readily learned to do. He then points out, however, that they could have done this on the basis of brightness differences among the papers, rather than color differences. He, therefore, equated each colored paper to a gray by eye, using a flicker technique. Next he tested each color against four equal grays, which he reported the animals found to be very easy discriminations. The inadequacies of his methods for eliminating possible brightness cues are obvious: The brightness match was for his eye rather than for the animal under investigation; and it would have been better in a four-choice test to have used three different grays, introducing slight variations about the presumed equality point.

There have been numerous subsequent studies of macaque color vision. Those of Watson (1909), Shepard (1910), and Bierens de Haan (1925) are primarily of interest for historical reasons and to reassure us that some progress has been made even in this field. Shepard (1910), for instance, in a step backwards from Kinnaman, used dishes of rice which he either left plain white, or soaked in quinine together with Congo red dye, and had the animal choose between the two dishes. Not only was there no control for brightness, but none for odor either. In fact, the quinine-soaked red stimulus was also wet while the white was dry!

More recent studies (Trendelenburg & Schmidt, 1930: Grether, 1939; DeValois, 1965; DeValois & Jacobs, 1968; DeValois, Morgan, Mead, & Hull, in press) have gone beyond the question of the mere presence of color vision posed in the earlier experiments and have entered into detailed examinations of the characteristics of macaque vision. Trendelenburg and Schmidt (1930) showed that the three macaques they tested could discriminate 589 nm (yellow) and 520 nm (green) light from white light, independent of intensity. This certainly indicates the presence of color vision, but is insufficient to establish whether it is dichromatic or trichromatic. Grether (1939) found that his one macaque discriminated each of a series of wavelengths, differing in 10 nm steps from 600 to 480 nm, from white with 100% accuracy. DeValois *et al.* (in press) got the same result with a group of five macaques. Thus, the macaque is clearly a trichromat.

The question of whether the macaque is a normal trichromat was explored in the anomaloscope test by both Trendelenburg and Schmidt and by DeValois *et al.* In the former experiment the macaque had to discriminate three different mixtures of light of 671 (red) and 535 nm (green) from a 589 nm (orange) light. The three mixtures were matched by a normal human observer with wavelengths of 580, 589, and 596 nm, respectively. Only the mixture that the humans matched with 589 nm was confused by the monkey with 589 nm monochromatic light. DeValois *et al.* (in press) tested which mixture of 646 (red) and 545 nm (green) matched 576 nm (yellow) for five macaques, with the

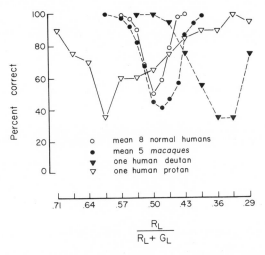

FIG. 3. Results of the anomaloscope tests of macaques and normal and anomalous human observers. The abscissa is the proportion of red in the red–green mixture. It can be seen that the protan requires more red, and the deutan more green to match a yellow than do the normal humans and macaques. (From DeValois, Morgan, Mead, & Hull, in press.)

results being compared with normal and anomalous human observers tested in the same apparatus. The animal was presented with the 646-nm light, plus three lights of 576 nm (whose brightnesses varied randomly about the equality point with the red). The animal had to choose the different light—in this case, the red. Then increasing amounts of the green light were added to the red. As this was continued in successive trials, the performance dropped from 100% correct discrimination to chance. As the addition of green (combined with a corresponding subtraction of red to keep the overall brightness constant) increased, the performance gradually rose again to 100% as the now greenish-looking mixture became distinguishable from the yellows. The data for the macaques and for the normal and anomalous human observers can be seen in Fig. 3, in which it is clear that the macaque is very similar to a normal trichromat and quite different from either prot- or deuteranomalous observers.

The complementary mixture of 610 (orange) and 495 nm (blue-green) was found for three macaques and three normal humans by Grether (1939). He found that the macaques required about twice as much orange in the orange–blue–green mixture as did the humans to match white. He concluded from this that the macaques are slightly protanopic, but this is contradicted by the anomaloscope data discussed above. As was suggested for similar data from the chimpanzee, a more likely explanation would be that the macaques, being young animals, have

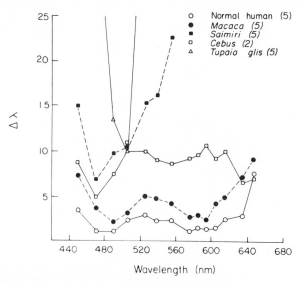

FIG. 4. Wavelength discrimination curves from different primate species tested in the same apparatus. *Tupaia* data from Polson (1968); *Cebus* from Morgan and DeValois (in preparation); human and *Macaca* from DeValois *et al.* (in press), and *Saimiri* from DeValois and Morgan (in press).

lower preretinal absorption in the short wavelengths and therefore higher sensitivity to blue, rather than lower sensitivity to orange.

Hue discrimination in the macaque has been studied by Trendelenburg and Schmidt (1930), Grether (1939), and by DeValois *et al.*, (in press). Trendelenburg and Schmidt (1930) tested the difference limen at 490 (blue), 535 (green), and 589 nm (orange) and found that both macaques and normal humans could discriminate smaller wavelength differences in the region of 490 and 589 nm than in that of 535 nm. Furthermore, the limens were at least as small for the macaques as for the humans. Grether (1939) tested at 500, 589, and 640 nm and found both the relative ordering of limens and the absolute hue discrimination ability of six macaques to be the same as his normal human subjects. DeValois *et al.* (in press) tested at 14 spectral points. These latter data are presented in Fig. 4. It can be seen that the shape of the curve is the same in these two species, although the normal human observers did somewhat better overall.

The extent to which macaque monkeys resemble normal human trichromats in both type and degree of color vision has been explored in another test, that of the relative saturation of the different parts of the spectrum, by DeValois *et al.* (in press). Five macaques and five human observers were tested in an

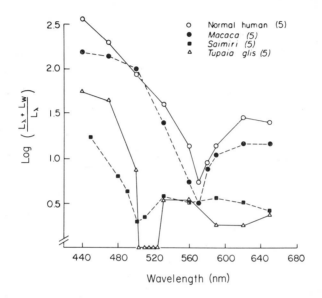

FIG. 5. Purity (saturation) discrimination curves from different primate species tested in the same apparatus. *Tupaia* data from Polson (1968); human and *Macaca* from DeValois and Morgan (in press); *Saimiri* from DeValois and Morgan (in press).

apparatus with four windows of which three were transilluminated with white light, and one with light from a monochromator. Responses to the colored window were rewarded. In successive trials, white light was added (and monochromatic light subtracted to keep brightness constant) to the monochromatic until the animal could no longer pick out the colored window. This was done at ten spectral loci. The results, presented in Fig. 5, show the excellent agreement between these two species both in the shape of the curve and in the absolute saturation of the spectrum.

Stiles (1959) showed that a flash of light of one color could be detected much more readily on a background of a different color than on one of the same color. This suggests that the adaptational changes in the various cone systems are to some extent independent of each other. Monjan (1966) has measured this selective adaptational process over a 4 log-unit range for two pigtailed macaques, two normal human trichromats, one deuteranomalous human, and one human protanope. He found quite similar amounts of masking of a red test flash by a red background relative to the effect of a green background with the macaques and normal human trichromats, and the same for green on green as compared with green on red. On the other hand, he found little or no effect with the color-defective human observers. Therefore, he concluded that macaques have the same red-green color vision system as do normal human trichromats.

Medin and Davis (1967) reported a study of color rating by macaques (mistitled a study of color discrimination, which it certainly is not). Groups of three trials were given. On the first, a pair of standard stimuli were presented and the animal had to choose, say, the left of two windows; then another pair of stimuli were shown and the animal had to choose the right window. Finally, on the third trial, another pair of stimuli intermediate between the first and second pair were shown, with a choice of either window rewarded. The point of interest was whether the animal chose the left window (thus presumably indicating that the third pair of stimuli were most like the first pair) or the right (indicating a greater similarity to the second pair). The stimuli used were Munsell chips varying in hue, brightness, and saturation. In some sets of trials the stimuli varied along only one of these dimensions; in others they varied along two, or along all three dimensions. They found that the monkeys made quite consistent ratings, with hue and brightness having large effects on the rating behavior, but saturation virtually none. When hue and brightness were paired, they added together in effectiveness.

As is well known, there is a wide variation in color vision among human males, at least 8% of whom differ sufficiently from the normal to be considered color defective. No one has tested a sufficiently large number of macaques (or any other nonhuman primate) to determine if a similar variation occurs in other primates.

In conclusion, the many studies of the macaque monkey show that it has highly developed color vision which appears to be virtually identical to that of normal man in both type and degree.

b. Other Species

The only information about color vision in Old World monkeys other than macaques comes from two animals with Grether (1939) included in his studies, and the amount of data he obtained was small. The difference limen of a vervet monkey *(Cercopithecus sabaeus)* was determined at 589 nm in the hue discrimination experiment; it was found to be similar to that of the macaque. A baboon *(Papio papio)* was examined in a bit more detail, on both the hue discrimination and the complementary-color test. It turned out to have color vision which, if anything, was closer to man than was that of the macaque.

It is highly unsatisfactory to draw conclusions about the nature of color vision in the wide variety of Old World monkeys other than the macques from such meager data, but it would certainly appear that Old World monkeys, in general, have excellent color vision, which closely resembles that of normal human trichromats.

3. CEBOIDEA

a. Cebus

Watson (1909), in one of the first studies of monkeys in the laboratory, found that the cebus monkey he tested could discriminate between red and green monochromatic lights, and between blue and yellow. However, as he recognized, his control of the intensity variable was insufficient to ensure that the animal was not choosing on the basis of brightness differences. Grether (1939) and Malmo and Grether (1947) examined four male cebus monkeys on a variety of color discrimination tests, finding clear evidence for the presence of color vision (but also showing that it is deviant from that of the normal human trichromat, as will be discussed later). For instance, three of the animals were found not only to be able to discriminate among different wavelengths when brightness was adequately controlled, but to have very small difference limens in certain spectral regions. One of the cebus monkeys was also found to be able to discriminate white from monochromatic light (the others were not tested on this problem).

Gunter, Feigenson, and Blakeslee (1965) studied one female and four male cebus monkeys in an experiment that reverted to the use of colored papers, instead of the monochromatic lights Trendelenburg and Schmidt and Grether had used two to three decades earlier. They also found evidence for color vision in these animals.

The first hint that cebus monkeys might be different from all of the Old World monkeys which had previously been studied, and which all turned out to resemble normal trichromats, came in Grether's (1939) measurement of hue discrimination limens at 500, 589, and 640 nm. The limen at 500 nm was about the same for the cebus monkeys as for macaques, but at 589 nm it was about twice as large, and at 640 nm about four times as large as the macaques'. Such decreased ability to discriminate among long wavelengths is not by itself definitive evidence for classifying an animal's color vision, since anomalous trichromats and dichromats of both the protan and deutan varieties have the same difficulty. The results did, however, strongly indicate some sort of color defect in these animals.

Gunter et al. (1965) also tested hue discrimination, among various yellow, orange, and red papers. The results are very difficult to relate to other studies of cebus or man, which have been carried out with monochromatic lights. They do indicate, however, the ability of cebus to make discriminations in this part of the spectrum and thus, to some extent, contraindicate a dichromacy of either the protan or deutan variety. Morgan and DeValois (in preparation) have also tested hue discrimination in two cebus monkeys, using the technique described in the discussion of the study by DeValois *et al.* of the macaque (Section III, 2, a). The animals were found to have a well-defined minimum in the hue-discrimination curve only at the short wavelengths, at about 500 nm. The discrimination among wavelengths above 530 nm was worse than that of the macaques, but superior to that of squirrel monkeys (see Fig. 4).

Since the hue discrimination data on his three cebus monkeys indicated a severe color defect, Grether (1939) tested for possible dichromacy by determining if there was a neutral point in the spectrum. By the time he ran this test, only one cebus was alive. Starting with 600 nm and going down the spectrum in 10-nm steps, Grether trained the animal to choose the mono-chromatic light from the white. The performance was 100% correct up to 530 nm, but fell to chance at 520 and 510 nm, and then reversed: the animal started to choose the white instead. The animal chose the white side on 100% of the trials from 500 to 480 nm. Thus, it appears that the region of 510–530 nm was a neutral band in the spectrum for this cebus, and Grether therefore concluded that these animals are dichromats. This conclusion was supported by the finding of a similar neutral point in another cebus (Malmo & Grether, 1947), although in this case, the animal could not discriminate wavelengths shorter than 500 nm from white either, which is peculiar and probably just reflects training problems.

Grether's views that cebus monkeys are dichromats gained wide currency, but recent evidence suggests that their color defect is probably not that extreme. Gunter *et al.* (1965) found that, while their five cebus monkeys had considerable difficulty discriminating some of the blue-green papers from gray ones (a difficulty that did not occur with other colors), the animals were in fact able to make the discriminations after extensive training. Morgan and DeValois (in preparation) obtained the same results with the discrimination of mono-chromatic lights from white: the cebus monkeys tested gave excellent performance everywhere except in the 480–510 nm range, where they initially fell to chance performance, just as Grether found. But with continued training the performance improved and the animals ended up showing clear evidence for discrimination even in this range. It is therefore much more reasonable to classify cebus monkeys as anomalous trichromats, rather than as dichromats.

Grether (1939) concluded that the color vision defect in cebus monkeys was in the protan direction mainly because they required about 270% more of the orange in a mixture of 495 and 610 nm to match white than did the normal humans. They therefore appeared to have a severe loss in sensitivity to the long wavelengths. If this were so, however, one would expect that they would have shown depressed sensitivity to the long wavelengths in their photopic sensitivity curve. Grether did not find this to be the case, but later Malmo and Grether (1947) did find a considerable loss in sensitivity to red in the one cebus monkey they tested. Some loss in sensitivity to long wavelengths was also found for three cebus monkeys by Morgan and DeValois (in preparation). The best test of whether the deviation from normal in the cebus monkey is in the protan or deutan direction would be the anomaloscope test, but as yet this has not been attempted. Another useful diagnostic test would be the determination of the isochromatic confusion lines. As was mentioned earlier, protanopes (and protanomalous trichromats to a lesser extent) confuse purples with blue, whereas deuteranopes confuse them with blue-greens and greens. Gunter *et al.* (1965) tested five cebus monkeys for this with Munsell papers and came to a rather complicated conclusion. Two animals made deutan mistakes, and the rest made some protan and some deutan errors. It would be very desirable to repeat this experiment with monochromatic lights and better experimental techniques, for much of the other evidence points strongly toward cebus monkeys being protanomalous trichromats.

The experiments on the cebus monkey have shown it to have color vision which is quite different from that of the Old World monkeys and normal man. The shape of the hue discrimination curve, the desaturation of the 480–500 nm spectral region, and the loss in long-wavelength sensitivity lead to the conclusion that the cebus monkey has color vision similar to that of human protanomalous trichromats.

b. Saimiri.

Klüver (1933), in a study that bore only incidentally on color vision, provided the first evidence for color vision in the squirrel monkey. Later, Miles (1958b) tested for wavelength discrimination in different spectral regions. Although Miles' intensity controls were not ideal—he made no attempt to equate the different wavelengths for brightness, but instead varied the intensities over a very large range—the results showed clear evidence for color vision.

Jacobs (1963) did a more extensive study of five squirrel monkeys, and found that they had color vision. In a neutral-point test with the white and monochromatic lights matched for brightness (on the basis of the monkey's photopic luminosity curve), these animals were able to discriminate among white and monochromatic lights in the long-wavelength part of the spectrum. However, the animals did show some initial confusion of white light and monochromatic lights below 520 nm, and, in fact, dropped to chance discrimination at about 490 nm. Thus these animals gave the initial appearance of being dichromats, but, after further training, could eventually distinguish all wavelengths from white. Since the part of the spectrum around 500 nm was examined at 2-nm intervals, it is clear that no true neutral point would have been missed. Even after extensive training, however, the region around 500 nm was only discriminated from the white with great difficulty, so this is clearly a highly desaturated region.

Further confirmation of the substantial desaturation of the 480–500 nm region for the squirrel monkey comes from studies of saturation discrimination by DeValois and Jacobs (1968) and DeValois and Morgan (in press). Using the same methods as those described earlier for the macaque and the cebus, they tested the saturation of ten wavelengths. The results of this test can be seen in Fig. 5, from which it is clear that the short wavelengths are much less saturated for the squirrel monkey than they are for normal human observers. The shape and height of the saturation function closely resemble that for protanomalous and deuteranomalous humans.

The results of hue discrimination tests on the squirrel monkey are consistent with the conclusion that they are anomalous trichromats. Miles (1958) found that the hue discrimination of squirrel monkeys was far worse than normal humans, although the wide variations in brightness that he used make the details of the function somewhat uncertain. He concluded that the animals were protanopes, like the cebus (which at that time were still thought to be dichromats), but one cannot diagnose dichromacy from hue discrimination curves alone. More recent studies of squirrel monkey hue discrimination (DeValois & Jacobs, 1968; DeValois & Morgan, in press) confirm the general shape of the curve found by Miles (see Fig. 4). These results show that *Saimiri* are even more deviant than cebus monkeys in this aspect of their vision.

Strong evidence that squirrel monkeys, like cebus monkeys, are protanomalous trichromats comes from still other tests (Jacobs, 1963). In the anomaloscope test to determine which mixture of 646 plus 545 nm matches 576 nm, it was found that the squirrel monkeys required far more red in the mixture than did macaques or normal human observers. This is the protan direction, and in fact the anomalous quotient (the green-to-red ratio required in the match for the subject divided by the same ratio for a normal observer) for the squirrel monkeys was 0.53, a value which was very close to that of a protanomalous human observer tested in the same apparatus. Still further evidence of the protan direction of the anomaly comes from the photopic luminosity curve of the squirrel monkey discussed earlier: Jacobs (1963) found these animals to be from 0.4 to 0.7 log units less sensitive to the long wavelengths than are macaque monkeys or normal human observers.

Thus, several lines of evidence lead to the conclusion that squirrel monkeys are severely protanomalous trichromats, with far poorer color vision than the Old World monkeys which have to date been studied. They also appear to be more deviant from normal trichromacy than are cebus monkeys, although both are protanomalous trichromats.

c. Other Species

None of the other New World monkeys has been examined nearly as carefully as cebus and squirrel monkeys. Grether (1939) included one spider monkey *(Ateles paniscus)* in his comprehensive study and found evidence for its having excellent color vision. The difference limens at 500, 589, and 640 nm were as low for this animal as for human observers given the same hue discrimination test. Unfortunately, no other tests were carried out on this animal. Glickman *et al.* (1965) also studied one spider monkey and found it to have color vision, but it had more trouble on color vision tests than one would have expected from Grether's results.

Two marmosets *(Callithrix)* were studied by Miles (1958a) for their ability to discriminate among blue, green, and red lights. Although the control of brightness is subject to the same criticisms offered earlier of the squirrel monkey study, the results strongly suggest that the marmoset has some color vision. Nothing is known, however, about the type or amount of color vision in this animal.

The only New World monkey—in fact, the only true monkey—that has a virtually all rod eye (containing only a few scattered cones) is the night or owl monkey *(Aotus trivirgatus)*. Ehrlich and Calvin (1967) failed to get any of four owl monkeys to make a red-green discrimination, although the monkeys were able to solve both brightness and form discrimination problems fairly readily. A failure to make a red-green discrimination by itself is, however, hardly sufficient evidence for monochromacy, since neither protanopes nor deuteranopes can discriminate between wavelengths in this range. Since this is the only reported study of color vision in the owl monkey, we must withhold judgment about its color vision.

4. PROSIMII

Galago. Most prosimians are nocturnal animals with all rod eyes, and would thus not be expected to have color vision (although that presupposition should be tested). Ehrlich and Calvin (1967) tested two galagos *(Galago crassicaudatus)* and found the same result as they did with the owl monkey, namely, an inability to discriminate between red and green. However, the same limitations on interpretation hold as just mentioned for their results with the owl monkey.

Lemur. An exception to prosimian nocturnality are the lemurs *(Lemur)*, which are crepuscular or diurnal in their activity cycle (Napier & Napier, 1967). The visual capabilities of these animals would be of great interest, but little information is available. Bieren de Haan and Frima (1930) attempted to study two mongoose lemurs *(Lemur mongoz)* and two ring-tailed lemurs *(Lemur catta)*, but the experiment was quite inconclusive. The ring-tailed lemurs were dropped because they would not respond in the apparatus, and the results of the color-vision tests on the other two were ambiguous. These lemurs eventually learned to distinguish red from blue, but showed no change in performance when certain grays were exchanged for the blue stimuli. Since there were inadequate brightness controls in the color tests, it is entirely possible that the whole discrimination was made on the basis of brightness. This possibility is further suggested by the fact that one of the animals could not discriminate among the colored and certain of the gray cards in another test. The other animal gave only feeble evidence of being able to make this same discrimination.

Glickman *et al.* (1965), in a study of one black lemur *(Lemur macaco),* found similar slight, but quite ambiguous, evidence for color vision. This animal, after a long training period, finally succeeded in discriminating between blue and red lights. No attempt was made, however, to measure the relative brightness of these two lights, and the random intensity variations introduced to eliminate brightness cues (a 1 log-unit range) may not have been sufficient for this purpose. This possibility is suggested by the results of a later test in which the colored light had to be discriminated from white, the luminance of which was varied over a 2 log-unit range. The performance dropped close to the chance level both for the red versus white and for the blue versus white.

The most reasonable conclusion is that lemurs might have color vision, but as yet this has not been demonstrated. In any case, it does not seem likely that they have highly developed color vision like the Old World monkeys and apes, or even like that of the New World monkeys.

Tupaia. While there is increasing doubt that the tree shrew should be classified as a primate, such has often been the case. Given their taxonomic position with respect to the primates, it is of interest in any case to examine the evidence with respect to their color vision. Tigges (1961, 1963a, 1964) studied several common tree shrews *(T. glis)* in a two-choice situation in which the animals had to discriminate a red, yellow, green, or blue card from each of 62 grays. They were able to do this with 85–100% accuracy at photopic levels, but not at scotopic levels. Furthermore, Tigges found that the animals could discriminate between various pairs of the colored cards, although the red–yellow and the green–yellow were hard for the animals. With adequate control of brightness, Tigges would have found these latter tests to be impossible for the tree shrews, for Polson (1968) has shown that these animals are, in fact, deuteranopic dichromats and, therefore, are unable to discriminate among lights of equal brightness in the green–yellow–red range of the spectrum.

After determining the photopic spectral sensitivity (and showing that there was no indication of scotopic function in this all-cone animal), Polson conducted neutral-point tests on five of the animals. They could easily discriminate all spectral colors from the white except in the region from 490 to 510 nm; when the monochromatic light was 505 nm, performance was at the chance level. Therefore, it appears that these animals are dichromats. Other color-vision tests to be described confirm this suspicion.

Polson (1968) also tested these tree shrews for wavelength discrimination at a number of spectral points. The results are shown in Fig. 4. Whereas these animals can discriminate among wavelengths in the blue-green portion of the spectrum about as well as can macaques or man, they are completely incapable of distinguishing among the long wavelengths. This is quite consistent with their being dichromats of either the protanopic or deuteranopic variety. The same conclusion can be drawn from the saturation discrimination results obtained by Polson and shown in Fig. 5. The short wavelengths are extremely desaturated for the tree shrew relative to the macaque or human, with the saturation around 500 nm being zero (their neutral point). Color vision in these animals is clearly very weak as well as being deviant.

That the dichromacy of the tree shrew is of the deuteranopic rather than the protanopic variety is indicated by two other experiments carried out by Polson (1968). First, she found that the photopic luminosity curve of this animal is not strikingly different from that of the macaque: The photopic function does not show the long-wavelength depression found in the protanomalous squirrel and cebus monkeys. Human deuteranopes show as high sensitivity to long wavelengths as do normals, their only sensitivity loss (if any) being in the middle of the spectrum. The other experiment which strongly suggests that tree shrews are deuteranopes is the test for isochromatic confusion lines. As mentioned earlier, deuteranopes confuse certain purples with greens or blue-greens, whereas protanopes confuse these same purples with blue. Polson (1968) showed that tree shrews do confuse purples with spectral colors, and that the colors with which they were confused were in the deuteranopic rather than the protanopic direction. In fact, the colors confused with the purples were slightly more toward the long wavelengths than were found even in human deuteranopes. These results all suggest that tree shrews are deuteranopic dichromats.

5. SUMMARY

Three different types of color vision have been found in the four main varieties of primates that have been extensively studied: apes, Old World monkeys, New World monkeys, and the tree shrew family of prosimians. The Old World monkeys, apes, and normal man, all appear to have essentially the same trichromatic color vision that we term "normal." Those of the New World monkeys that have been carefully examined are severely protanomalous. Finally, the tree shrews studied are dichromats of the deuteranopic variety. Of course, one finds all of these varieties of color vision in man alone. Whether or not other primates show the great intraspecies variation in color vision shown by man is not known.

IV. SPATIAL AND TEMPORAL ASPECTS OF VISION

A. Visual Acuity

Visual acuity is often defined as the visual angle subtended by the smallest detail that can be discriminated under a specific set of conditions. Of utmost importance in this definition is the phrase, "a specific set of conditions." As has been repeatedly pointed out in the past (LeGrand, 1967), there are as many different visual acuities as there are test situations, so statements about *the* visual acuity for a particular animal are simply not meaningful. This makes a comparative analysis especially difficult, because the test situations are rarely the same. The reader is urged to keep these considerations in mind in what follows.

The literature on visual acuity in man is voluminous, but the total number of such investigations in other primate species runs to fewer than two dozen articles. Because of the extensive information on human visual acuity, most of the discussion here will consider the results with nonhuman primates in the light of what is known of visual acuity in man.

1. Determinants of Visual Acuity

Acuity is dependent on several features of the physiology of the visual system (dioptric, retinal, and central nervous system), and on numerous features of the test situation. Examination of some of these determinants of visual acuity follows.

Three types of visual acuity tests have classically been studied. The first involves the discrimination of the presence of a test object, typically a single line or point (minimum visible test). This test shares many of the features of a brightness discrimination task and usually results in high visual acuity. There has been only one study published using this technique: Klüver (1933) found that a rhesus monkey was able to discriminate the presence of a black thread on a white background when the thread subtended only 21 seconds of arc. A second type of acuity test, measurement of vernier or offset acuity, has not been used with nonhuman primates. The third, and most popular test of visual acuity involves the determination of the minimum-separable visual angle necessary for discrimination; most of the studies of interest here are of this type.

Whereas older studies of visual acuity have considered only the minimum discriminable object, more recent investigations have expanded this areas to include an examination of the detectability of objects of all sizes. For example, if subjects are shown displays consisting of a series of parallel lines, with the lines varying in thickness from dispay to display, under conditions of maximum contrast (black on white) all but the finest lines in these grating patterns will be seen as separate. The usual measure of acuity in this case is the narrowest line (highest spatial frequency) which can be resolved. However, if the contrast is reduced (dark gray on light gray instead of black on white) larger lines will not be resolvable. The minimum contrast required for resolution can thus be used to measure acuity for any object size. Such a study (often called a modulation transfer investigation) is more informative than merely measuring acuity at 100% modulation (black on white). In the case of the human visual system, for instance, it turns out that large test objects (low spatial frequencies) cannot be resolved nearly so well as they should be on purely optical grounds, implicating the presence of neural spatial inhibitory mechanisms. It would certainly be most useful to know whether similar low spatial-frequency losses are found in other primates.

The majority of studies of minimum-separable visual acuity have utilized a grating pattern as the test object. Typically a pair of diffraction gratings are mounted one in front of the other and then rotated in opposite directions (Johnson, 1914; Spence, 1934; Weinstein & Grether, 1940; Weiskrantz & Cowey, 1963, 1967; Cowey & Ellis, 1967, 1969). When diffuse light passes through such pairs diffraction patterns are created, with the widths of the resultant Moiré fringes dependent on the angle of rotation between the gratings. In each of these investigations, the animal was initially trained to discriminate a broadly striated field from a homogeneous field, then the widths of the stripes were reduced until the discrimination fell to chance. Other investigators (Woodburne, 1965; Treff, 1967; Ordy & Samorajski, 1968) have used grating patterns as test stimuli, but required their subjects to discriminate between two identical gratings differing only in their orientations. It is unlikely that these two procedures (homogeneous versus striped field or striped versus striped field) would lead to the same results.

Other investigations have assessed minimum-separable visual acuity by using the familiar Landolt-C pattern where the minimum discriminable gap in the letter is determined. Farrer and Graham (1967), using the Landolt pattern in a four-alternative, oddity problem, trained a rhesus monkey to select the C whose gap-orientation differed from the other three stimuli. Yarczower, Wolbarsht, Galloway, Fligsten, and Malcolm (1966) trained a stumptailed macaque to indicate by the nature of its response whether the gap was to the right or to the left. Again, it is unlikely that measurements of minimum-separable visual acuity by grating and by Landolt pattern would yield comparable acuity values.

In cases where training procedures are difficult, or impossible, minimum-separable acuity has been assessed by recording the optokinetic nystagmus response. The general procedure in such studies (Ordy, Massopust, & Wolin, 1962; Ordy, Latanich, Samorajski, & Massopust, 1964; Reisen, Ramsay, & Wilson, 1964) has been to utilize a drum which rotates a striated black-white pattern in front of the subject's eyes: The angle subtended by the narrowest stripes that regularly elicit a rhythmic optokinetic response is taken as representing acuity threshold.

It has been well established that target luminance influences human visual acuity. Only two studies have examined the effect of luminance on acuity for nonhuman primates. Spence (1934), in his measurements of visual acuity of two chimpanzees, examined several luminance levels between 0.00138 and 28.25 mL. He found visual acuity to improve from about 0.90' of arc to 0.44' of arc as luminance level increased, the improvement being about the same for both chimpanzee and human subjects. Similarly, Weinstein and Grether (1940) obtained acuity measurements from two rhesus monkeys at three luminance levels and also found an improvement in visual acuity that paralleled the improvement shown by human observers in the same test situation. Other studies of visual acuity have not shown an appreciation for the importance of the luminance of the test target. For instance, the luminance has varied as much as 2 log-units across studies that have sought to measure acuity in the same species. In other studies, target luminance level is not even reported.

For brief stimulus durations human visual acuity is known to follow a time-intensity reciprocity law, whereby reductions in stimulus duration can be offset by increases in stimulus intensity (Graham & Cook, 1937). It would be of considerable interest to know if, and under what conditions, this reciprocal relationship also holds for other primate species. As yet, there are no data on this issue. In fact, only occasionally do any of these studies mention any restriction on viewing time: Typically the occurrence of the animal's response, rather than the experimental protocol, regulates exposure duration. The technique of making the animal perform an observing response in order to initiate the test trial (Yarczower *et al.*, 1966) should prove useful to experimenters interested in controlling stimulus exposure duration.

Other things being equal, visual acuity should not change as a function of the distance of the test target from the eye. Spence's (1934) finding that visual acuity was superior for two chimpanzees when the test gratings were presented at a distance of 120 cm than when they were presented at 75 cm is likely due to focusing problems and accommodation fluctuations, either of which may effectively reduce visual acuity (Westheimer, 1965).

Orientation of the test target as a determinant of visual acuity has not as yet been systematically investigated, most of the test gratings having been presented in the vertical or horizontal position. In humans, acuity is known to be better for horizontal and vertical test objects than for similar test objects oriented obliquely (Taylor, 1963). Since cortical cells of different orientation specificity have been found (Hubel & Wiesel, 1962), there exists a plausible physiological basis for such a horizontal–vertical superiority. It is important to know if the results for humans hold for those nonhuman primates that can be used for the relevant physiological research. It has been shown that at least some monkey species are able to detect the displacement of a line from the vertical with about the same sensitivity as human observers. Payton and Blake (1964) found that rhesus and cynomolgus macaques and vervets *(Cercopithecus aethiops)* could successfully discriminate a line deviating by only 1° from the vertical. Whether there were any differences in this ability among these three monkey species is, unfortunately, not reported.

Nearly all measurements of visual acuity have been accomplished under binocular viewing conditions. One study on a rhesus monkey examined both monocular and binocular viewing by making use of a face mask for the monkey (Farrer & Graham, 1967). Acuity thresholds were between 0.5 and 1.0' of arc for both monocular and binocular viewing but there is a hint of binocular superiority: For a stimulus which subtended 1' of arc, monocular viewing yielded about 75% correct discrimination and binocular viewing led to 90% correct. Thus binocular vision may yield superior visual acuity for the rhesus monkey, just as it does for the human observer (Campbell & Green, 1965).

Human visual acuity declines systematically as the test stimulus is moved from the fovea to the periphery and that decrease follows (but only approximately) the decline in cone density. It is unknown how general this phenomenon is among the various other primate species. As yet, there are no studies in the literature on nonhuman primates, which have determined visual acuity at various retinal loci. Instead, the general procedure has been to allow the animal subject ot determine the locus of stimulation, the assumption being that he will choose the one that will maximize acuity. A secondary assumption is that such a procedure will lead to an estimate of foveal acuity. Procedures for controlling the locus of retinal stimulation are, however, available. For example, Cowey (1963) reported perimetric measurements for rhesus monkeys obtained by photographing corneal reflections and then assessing *post hoc* the direction of the gaze and, hence, the retinal locus of visual stimulation. Such a procedure could also presumably be utilized for mapping visual acuity at various retinal locations.

Two studies have investigated the effects on visual acuity of destroying the central portions of the retina. The postlesion results provide some information about peripheral acuity. Weiskrantz and Cowey (1963) found a substantial loss of acuity in rhesus monkeys following foveal lesions. The relation of lesion size to the magnitude of loss in visual acuity led them to suggest that the acuity fall-off in the periphery might be somewhat smaller for the rhesus monkey than for the human observer. Similarly, Yarczower *et al.* (1966) found that following macular destruction of both retinas of a stumptailed macaque, visual acuity deteriorated to 9.0' of arc from a preoperative value of 1.4' of arc.

The presence of a retinal fovea is usually thought to be associated with high visual acuity. Since there are numerous structural variations in retinas of different primate species, a comparative correlation of visual acuity with retinal morphology would seem to be of considerable interest. A recent study by Ordy and Samorajski (1968) made such comparisons of visual acuity and retinal structure across six primate species. Minimum-separable acuity for grating test patterns was found to be 0.5–1.5' of arc for squirrel monkeys and marmosets *(Callithrix jacchus),* both of which have duplex retinas and a well-defined fovea. Owl monkeys and galagos *(Galago crassicaudatus)* both have nearly all-rod retinas and an *area centralis* but no fovea. Their visual acuities were found to be from 3.5 to 9.0′ of arc. Most interesting in this study are the results from common tree shrews and from ring-tailed lemurs. The visual acuity of each of these species is comparable to that for squirrel monkeys and marmosets (0.5–1.5′), although neither the tree shrew nor the lemur has a clear fovea. The retina of the tree shrew contains predominantly cones with an "*area centralis*" located far out on the temporal retina (Rohen & Castenholtz, 1967). The lemur retina, on the other hand, contains predominantly rods, although it also shows a distinct *area centralis*. Ordy and Samorajski suggest that in their comparative survey, the most consistent anatomical feature associated with high visual acuity is a small convergence of receptors onto ganglion cells. It would be useful to see this sort of inquiry expanded to other species, and with more precise measurements of visual acuity. For example, Polyak (1946) has described the mangabey as possessing the highest cone density among the primates; this monkey should thus be included in a correlative study of retinal morphology and visual acuity.

For some comparisons, it is necessary to translate into retinal dimensions acuity measurements made in terms of visual angle. Since eye size varies considerably among primate species, specification of visual angle does not specify the distance on the retina. This latter computation requires a knowledge of the posterior nodal distance of the eye. To date, most studies have only reported measures of visual angle. One of the few experiments in which visual acuity thresholds were specified in terms of retinal image width for several primate species was that of Grether (1941b). His measurements demonstrated that, although human visual acuity is superior to that of the rhesus monkey in terms of angular measurements, the retinal image widths for threshold acuity are closely similar for the two species. He also calculated retinal image widths from the results of other studies and concluded that acuity, in terms of image width, is about the same for man, chimpanzee, and rhesus monkey. Cowey and Ellis (1967) found that, although visual acuity is slightly better for the rhesus than for the squirrel monkey in terms of visual angle, the considerably smaller eye of the squirrel monkey must actually be resolving a smaller retinal image at discrimination threshold. Further measurements of this kind would provide a needed supplement to the typical reports on acuity given solely in terms of angular specification.

That a large proportion of humans require glasses for optimal visual acuity is not exactly an obscure fact. Those who have studied visual acuity in nonhuman primates seem not to have considered that the same might be true for their animals. Young (1963, 1964) has refracted large numbers of macaques and chimpanzees and has found a distribution of refractive errors that is similar to those found in man. It is clear, therefore, that an experimenter picking an animal or two of some species to study might very well choose a very myopic or presbyopic one by chance. He would thereby get results that were quite unrepresentative of the species as a whole, or of the "neural" resolving power of even that eye itself.

2. ACUITY AS A VISUAL INDEX

One might expect that since visual acuity is straightforward in concept and relatively easy to measure, it would be widely used as an index of vision in studies of other aspects of behavior. So far, such investigations utilizing nonhuman primates have been few in number and have dealt either with the development of vision, or the effects of lesions on visual discrimination.

Ordy and co-workers (Ordy, Massopust, & Wolin, 1962; Ordy, Latanich, Samorajski, & Massopust, 1964) measured the ontogenetic development of visual acuity in several primate species. An orangutan, a gibbon, three baboons, four rhesus monkeys, and 44 humans infants were tested for the presence of an optokinetic nystagmus response at birth, and over the first two months of development. For newborns, minimum separable acuity ranged from a low of 40' for the humans to a high of 9' for the baboons. Following birth, there was a rapid improvement in visual acuity among all primates, with the fastest increase shown by the baboons, followed in order by the rhesus monkeys, the gibbon, the orangutan, and the humans. The differences in development rate between man and the nonhuman primates were particularly marked. By the age of 9 weeks all nonhuman primates had reached a threshold of 4', a level that the human does not achieve until about 6 months of age. Ordy et al. (1962) also report that the rhesus monkey can resolve separations as small as 1' by the end of 2 months whereas only about 52% of human children achieve this level by 7 years of age.

In another developmental study, Riesen, Ramsey, and Wilson (1964) assessed the effects of early deprivation of patterned light on the subsequent development of acuity in rhesus monkeys. After being raised for the first 20 or 60 days of life in diffuse light, these monkeys received 2.5 hours/day of patterned light experience. They found that by the end of 30 days of this patterned light exposure, the deprived animals had achieved the same level of visual acuity as normal monkeys do at the same age.

Weiskrantz and Cowey (Weiskrantz & Cowey, 1963; Weiskrantz & Cowey, 1967; Cowey, 1967; Blakemore, Hodkinson, & Cowey, 1968; Cowey & Ellis, 1969) have utilized measurements of visual capacity to evaluate the effects of structural damage of the visual system, particularly the relationships between retinal and cortical lesions. Behavioral measurements utilized in these experiments include both measures of grating acuity and the use of the perimetric technique mentioned above. In several careful experiments, these investigators have established that the scotoma produced by lesions in visual cortex is not absolute, even when the entire occipital lobe is removed. Possible explanations are that extensive lateral spread of information at the retinal level might allow activity arising from the center of the scotomatous region to activate an intact portion of the cortex, or that the discrimination might be based on retinal information that goes to the superior colliculus or other subcortical terminations. These alternatives cannot as yet be discriminated. One of the complicating factors is that in these experiments and other lesion studies, one often finds improvement in visual sensitivity following the loss, which may or may not be related to practice effects (Cowey, 1967).

Similarly, losses in visual acuity following ablation of the visual cortex are smaller than would be expected if cortical destruction produced a complete scotoma. Weiskrantz and Cowey (1967) made a direct comparison of the acuity losses engendered by cortical and retinal lesions. They found that retinal lesions led to larger acuity reductions than did cortical lesions although the lesions were comparable in the size of the visual field defects they produced. Again, they invoke lateral interaction effects to explain the disparate results of these two types of destruction.

Blakemore *et al.* (1968) have provided direct evidence on the recovery of visual function following lesion. They examined the frequency and extent of misreaching for food by rhesus monkeys after foveal destruction. The magnitude of the reaching error, and the rate of recovery from misreaching were directly related to the size of the macular lesion. Complete disappearance of misreaching requires only 12–30 days following the operation and they suggest that a functional reorganization of visual processing allows information from the intact, peripheral retina to be utilized for performance of this task. That the recovery involves both afferent and efferent changes is indicated by the observation that part of the reaching improvement is due to a change in the animal's behavioral strategy—during recovery a monkey begins to spread and use its whole hand for grasping, rather than just the thumb and forefinger as do nonlesioned animals.

3. COMPARATIVE SURVEY OF VISUAL ACUITY

As indicated at the beginning of this section, at present no truly comparative survey of visual acuity among nonhuman primates is possible. Nevertheless it might be useful to bring together the available information on visual acuity. Table II presents such information, indicating the nature of the test utilized as well as the measure of visual acuity. The figures given in the table represent either the mean or the range of acuities obtained. In general, the differences in acuity found for the different species tested in the same experiment, or in different experiments by the same investigators, are more likely to be valid than other comparisons. Finally, at the risk of being unduly repetitive, we should note again that many of the variables known to be important determiners of

TABLE II

Visual Acuity Measurements in Nonhuman Primates

Genus	Test type[a]	Visual acuity[b]	References
Pan	Grating, S vs H	0.47	Spence (1934)
Macaca	Landolt C	0.5–1.0	Farrer & Graham (1967)
Macaca	Grating, S vs H	0.60	Weiskrantz & Cowey (1967)
Macaca	Grating, S vs H	0.65	Cowey & Ellis (1967)
Macaca	Landolt C	1.4	Yarczower *et al.* (1966)
Macaca	Grating, S vs H	0.57–1.09	Weiskrantz & Cowey (1963)
Macaca	Grating, S vs H	0.67	Weinstein & Grether (1940)
Macaca	Minimum visible	0.13	Klüver (1933)
Cebus	Grating, S vs H	0.95	Johnson (1914)
Saimiri	Grating, Hor vs V	0.5–1.5	Ordy & Samorajski (1968)
Saimiri	Grating, S vs H	0.74	Cowey & Ellis (1967)
Saimiri	Grating, Hor vs V	0.84	Woodburne (1965)
Aotus	Grating, Hor vs V	3.5–8.0	Ordy & Samorajski (1968)
Callithrix	Grating, Hor vs V	0.5–1.5	Ordy & Samorajski (1968)
Galago	Grating, Hor vs V	3.5–8.0	Ordy & Samorajski (1968)
Galago	Grating, Hor vs V	4.28	Treff (1967)
Lemur	Grating, Hor vs V	0.5–1.5	Ordy & Samorajski (1968)
Tupaia	Grating, Hor vs V	0.5–1.5	Ordy & Samorajski (1968)

[a] S = striped field, H = homogeneous field, Hor = horizontal orientation, V = vertical orientation.

[b] Stated as the visual angle, measured in minutes of arc.

visual acuity for human observers (viewing time, retinal location, luminance, etc.) have been uninvestigated, uncontrolled, or ignored in these experiments.

B. Depth, Form, and Movement

There is a startling paucity of information on depth, form and movement vision in nonhuman primates and certainly nothing that provides the basis for a comparative survey. These features have, of course, appeared as discriminative cues in may experimental situations; in particular, stimulus shape has been widely used as the discriminative stimulus in numerous learning studies. These latter studies provide little or no information on visual capacity *per se,* and will not be considered here.

Discrimination of visual depth can be demonstrated at a very early stage of development in the rhesus monkey as in the human. Rosenblum and Cross (1963) found that a group of infant rhesus monkeys tested in a visual-cliff situation gave clear evidence for depth discrimination as early as the third day of life. Two other recent studies report measurements of binocular depth discrimination. Treff (1967) showed that the two galagos he tested could discriminate successfully a depth difference amounting to about 40' of visual angle, considerably poorer than a human observer in the same situation. Predictably, he also found that the accuracy of depth perception for the galago is affected by such features of the experimental situation as the physical arrangement of the discriminanda, and the illumination level at which the task is performed. That optimal depth perception depends on the full use of both eyes in this species was shown by the decrement in performance that resulted from paralysis of accommodation in one eye. In a somewhat similar study, Schmidt (1968) found the threshold for depth discrimination to be about the same for squirrel monkeys and man, just as was true for visual acuity. The squirrel monkey's depth perception was also found to be independent of the tilt in head position. From observations of this primate in a jumping task, Schmidt suggests that the squirrel monkey strongly depends on motion parallax for visual orientation.

In a study mentioned above, Payton and Blake (1964) reported that macaques and grivets were able to discriminate the displacement of a line from the vertical with a sensitivity approaching that of human observers. A later study (Reid, Medin, & Davis, 1965) provides verification of this result for both rhesus and pigtailed macaques. In addition, Reid et al. made some observations on the cues utilized by monkeys in the perception of verticality. They found that monkeys, like humans, prefer visual over postural cues when both are available, but could use either or both simultaneously in particular situations. No differences between species, nor between jungle and laboratory-reared animals were noted in these tests.

A most important aspect of normal vision is the ability of the individual to recognize objects despite many types of changes in the retinal image. This is referred to as "perceptual constancy." Varied and complex conditions lead to its presence or absence in human observers. Early observations by Locke (1938) and by Köhler (1915) suggested the occurrence of size constancy in monkeys and apes. More recently, Zeigler and Leibowitz (1958) designed an experiment to assess the occurrence of shape constancy in monkeys. First they trained rhesus monkeys on an area discrimination task and then tested for constancy by opposing the area and the shape of the stimulus. The rhesus subjects followed much more closely the results predicted on the basis of constancy than did human observers tested on the same task. Thus, at least in this one situation, monkey subjects are capable of maintaining perceptual constancy in the face of large changes in the retinal stimulus.

Although both studies of optokinetic nystagmus, mentioned previously, and studies of temporal resolution surveyed below, provide some information relevant to an understanding of movement perception, there is apparently only a single directly relevant study. Carpenter and Carpenter (1958) measured thresholds for the perception of movement in young chimpanzees and human children when all subjects were at about the same maturational level. Thresholds for movement in both these species were about the same—on the order of 25 to 100 min of visual angle per sec as compared to the 10 to 17-min figure achieved by adult human observers in the same experimental setting. In view of the current extensive interest in the physiological correlates of movement perception, many more studies of this aspect of primate perception would be welcome.

C. Time as a Stimulus Parameter

It has already been pointed out that in most experiments reviewed in this section, very little attention has been paid to the duration of the observation interval. An examination of the literature in which there has been a more explicit interest in time as a stimulus parameter reveals some surprising omissions. For example, the time-luminance reciprocity, formalized as Bloch's law, which has been the subject of numerous experiments with human observers, has, apparently, not been investigated in any other primate species. Similarly, temporal changes in sensitivity and the multitude of parameters on which these changes depend, have received little attention although the use of the Békésy tracking technique with monkey subjects provides one useful method for assessing these time changes. Experiments utilizing time as a stimulus parameter will be presented in three categories.

1. TEST-FLASH DURATION

We consider here the minimum exposure duration necessary for pattern discrimination. Good discrimination in such situations depends on such features as distance from discriminanda to observer, the contrast relation between the discriminative stimulus and the visual background, and the nature of the target to be discriminated, in addition to the exposure duration. Pribram, Spinelli, and Kamback (1967) report that monkeys are capable of making successive pattern discriminations with test-flash durations as short as 0.01 msec. Adkins, Fehmi and Lindsley (1969) trained animals in a simultaneous discrimination problem and found that a pigtailed macaque could discriminate accurately between a square and a triangle at exposure durations of 0.5 msec. This latter study incorporated an observing-response procedure which probably facilitates discrimination for brief test flashes. Both of these studies suggest that the minimal stimulus duration for the macaque approaches that for man in comparable discrimination situations.

2. CRITICAL FLICKER FREQUENCY

The rate of stimulus intermittence at which the appearance of a light changes from a flickering one to a steady one (the critical flicker frequency or CFF) has been the subject of a very large number of studies with human observers, and several with other primates. The CFF has been measured by several different methods. Symmes (1962) trained rhesus monkeys with a continuous tracking procedure in which the animal pressed one lever to raise the frequency of an intermittent light and a second to lower its frequency, the reinforcement contingencies being so arranged that the animal adjusts frequency about the CFF point. More frequently, investigators have employed some sort of a discrete trials task, one variant of which is a go–no-go procedure in which the animal learns to respond in the presence of a flickering light stimulus and to withhold its response in its absence (Brecher, 1935; Mishkin & Weiskrantz, 1959). The discrete trials procedure has also been used with forced-choice techniques on several occasions, and with a variety of primate species (Jacobs, 1963; Adams & Jones, 1967; Polson, 1968; DeValois et al., in press). In these experiments the animal was presented with multiple alternatives and had to select which stimulus from among an array appeared to be flickering. Still another approach has recently been used in which a conditioned-suppression procedure is exploited (Shumake, Smith, & Taylor, 1968). These investigators first trained three rhesus monkeys to press a lever on a variable-interval reinforcement schedule. Following acquisition of this response, the animals received 30-sec episodes of flicker each of which terminated in an unavoidable foot shock. The magnitude of response suppression during the preshock period was used as an index of discrimination.

Given the variety of experimental situations, and the traditional lack of reports on inter- and intra-subject variability, only the most casual comments can be made on the relative utility of these various techniques for CFF measurement. Symmes (1962) reported that monkeys tested with his tracking procedure yielded slightly lower CFF values than did animals tested with the go–no go procedures under the same stimulus parameters. The forced-choice procedures probably lead to faster acquisition of the task and have the advantage that, with multiple stimulus alternatives, the animal learns an oddity response rather than an absolute stimulus discrimination. The conditioned-suppression procedure, as utilized by Shumake et al., leads to quite small inter-subject variability and also seems to provide a reliable CFF value (retest variability of only 0–5 Hz). At the present no single best method can be recommended for CFF measurement in monkeys.

One justification for the determination of CFF in nonhuman primates has been to provide a visual response measure with which the effects of central nervous system damage may be evaluated. As examples of this approach, work by Mishkin and Weiskrantz (1959) and by Symmes (1965) may be briefly mentioned. These investigators determined the CFF value before and after cortical lesion. Very little change in final CFF values resulted from a variety of such lesions even though extensive retraining was sometimes required. Mishkin and Weiskrantz noted that steady improvement in CFF performance could be discerned for up to one year. Symmes found that a loss in CFF level depends more critically on the size of the cortical lesion than on its location.

The CFF measurement has also been used as an index of visual capacity. For example, since rod and cone receptors have different time constants, CFF data can be used to detect the presence of a Purkinje shift and to obtain separate scotopic and photopic luminosity functions. Both of these uses of CFF data have been reviewed in Section II. Here only a few additional observations of comparative interest will be added.

In an early study, Brecher (1935) found the CFF for a rhesus monkey to increase linearly with log intensity over three log units of luminance (to a maximum of 17 mL). Over this range, the CFF for the monkey subject varied from 9 to 31 cps and the human subjects from 13 to 38 cps. A similarly close correspondence between man and macaque monkey has been reported by DeValois *et al.* (in press); they reported that the luminance necessary for CFF over a range of 10 to 35 cps never varied by more than 0.5 log units among human and macaque subjects run under the same procedures.

3. SEQUENTIAL EFFECTS

Recently much interest has been centered on the experimental fact that a visual stimulus may retroactively interfere with the detection of another visual stimulus. One recent study clearly demonstrates that this backward masking effect occurs in primates other than man. Adkins *et al.* (1969) first trained pigtailed and stumptailed macaques on a form discrimination. After that discrimination had been acquired, a masking flash of diffuse light was paired with the test stimuli, the interval between the test flash and the masking flash being the experimental variable. If the masking flash came 50 msec or more after the test stimulus it had no effect. If, however, the interflash interval was reduced to about 30 msec, discrimination performance began to deteriorate, and at about a 20-msec separation the interference was complete. These time relationships are very close to those found for humans and thus this experiment provides both another case where human and macaque visual capacities seem to be highly similar, and a paradigm for the investigation of complex time effects with monkey subjects.

REFERENCES

Adams, C. K., & Jones, A. E. Spectral sensitivity of the sooty mangabey. *Perception & Psychophysics,* 1967, **2,** 419-422.
Adkins, J. W., Fehmi, L. G., & Lindsley, D. B. Perceptual discrimination in monkeys: Retroactive visual masking. *Physiology & Behavior,* 1969, **4,** 255-259.

Bierens de Haan, J. A. Experiments on vision in monkeys I. The colour-sense of the pig-tailed macaque (*Nemestrinus nemestrinus* L.). *Journal of Comparative Psychology*, 1925, 5, 417-453.

Bierens de Haan, J. A, & Frima, M. J. Versuche über den Farbensinn der Lemuren. *Zeitschrift für vergleichende Physiologie*, 1930, 12, 603-631.

Blakemore, C., Hodkinson, R. G., & Cowey, A. Retinal lesions in monkeys: Recovering from misreaching. *Vision Research*, 1968, 8, 883-888.

Blough, D. S., & Schrier, A. M. Scotopic spectral sensitivity in the monkey. *Science*, 1963, 139, 493-494.

Brecher, G. A. Die Verschmelzungsgrenze von Lichtreizen beim Affen. *Zeitschrift für vergleichende Physiologie*, 1935, 22, 539-547.

Brooks, B. A. Neurophysiological correlates of brightness discrimination in the lateral geniculate nucleus of the squirrel monkey. *Experimental Brain Research*, 1966, 2, 1-17.

Campbell, F. W., & Green, D. G. Monocular versus binocular visual acuity. *Nature*, 1965, 208, 191-192.

Carpenter, B., & Carpenter, J. T. The perception of movement by young chimpanzees and human children. *Journal of Comparative & Physiological Psychology*, 1958, 51, 782-784.

Cowey, A. The basis of a method of perimetry with monkeys. *Quarterly Journal of Experimental Psychology*, 1963, 15, 81-90.

Cowey, A. Perimetric study of field defects in monkeys after cortical and retinal ablations. *Quarterly Journal of Experimental Psychology*, 1967, 19, 232-245.

Cowey, A., & Ellis, C. M. Visual acuity of rhesus and squirrel monkeys. *Journal of Comparative & Physiological Psychology*, 1967, 64, 80-84.

Cowey, A., & Ellis, C. M. The cortical representation of the retina in squirrel and rhesus monkeys and its relation to visual acuity. *Experimental Neurology*, 1969, 24, 374-385.

Crawford, M. P. Brightness discrimination in the rhesus monkey. *General Psychological Monographs*, 1935, 17, 71-161.

Dartnall, H. J. A., Arden, G. B., Ikeda, H., Luck, C. P., Rosenberg, C. M., Pedler, C. M. H., & Tansley, K. Anatomical, electrophysiological and pigmentary aspects of vision in the bush baby: An interpretative study. *Vision Research*, 1965, 5, 399-424.

DeValois, R. L. Behavioral and electrophysiological studies of primate vision. *In* W. D. Neff (ed.), *Contributions to sensory physiology*, Vol. 1. New York: Academic Press, 1965. Pp. 137-178.

DeValois, R. L., & Jacobs, G. H. Primate color vision. *Science*, 1968, 162, 533-540.

DeValois, R. L., & Morgan, H. C. Psychophysical studies of monkey vision: III. Squirrel monkey wavelength and saturation discrimination. *Vision Research*, in press.

DeValois, R. L., Abramov, I., & Jacobs, G. H. Analysis of response patterns of LGN cells. *Journal of the Optical Society of America*, 1966, 56, 966-977.

DeValois, R. L., Polson, M. C., & Morgan, H. C. Psychophysical studies of monkey vision: I. Methods and macaque luminosity tests. *Vision Research*, in press.

DeValois, R. L., Morgan, H. C., Mead, W. R., & Hull, E. M. Psychophysical studies of monkey vision: II. Macaque color vision tests. *Vision Research*, in press.

Ehrlich, A., & Calvin, W. H. Visual discrimination behavior in galago and owl monkey. *Psychonomic Science*, 1967, 9, 509-510.

Farrer, D. N., & Graham, E. S. Visual acuity in monkeys: A monocular and binocular subjective technique. *Vision Research*, 1967, 7, 743-747.

Glickman, S. E., Clayton, K., Schiff, B., Guritz, D., & Messe, L. Discrimination learning in some primitive mammals. *Journal of Genetic Psychology*, 1965, 106, 325-335.

Graham, C. H. *Vision and Visual Perception.* New York: Wiley, 1965.

Graham, C. H., & Cook, C. Visual acuity as a function of intensity and exposure time. *American Journal of Psychology*, 1937, 49, 654-661.

Grether, W. F. Color vision and color blindness in monkeys. *Comparative Psychology Monographs*, 1939, **15**, No. 4. 1-38 (whole No. 76).

Grether, W. F. Chimpanzee color vision. I. Hue discrimination at 3 spectral points. *Journal of Comparative Psychology*, 1940, **29**, 167-177. (a)

Grether, W. F. Chimpanzee color vision. II. Color mixture proportions. *Journal of Comparative Psychology*, 1940, **29**, 179-186. (b)

Grether, W. F. Chimpanzee color vision. III. Spectral limits. *Journal of Comparative Psychology*, 1940, **29**, 187-192. (c)

Grether, W. F. A comparison of human and chimpanzee spectral hue discrimination curves. *Journal of Experimental Psychology*, 1940, **26**, 394-403. (d)

Grether, W. F. Spectral saturation curves for chimpanzee and man. *Journal of Experimental Psychology*, 1941, **28**, 419-427. (a)

Grether, W. F. Comparative visual acuity in terms of retinal image widths. *Journal of Comparative Psychology*, 1941, **31**, 23-33. (b)

Grether, W. F. The magnitude of simultaneous color contrast and simultaneous brightness contrast for chimpanzee and man. *Journal of Experimental Psychology*, 1942, **30**, 69-83.

Gross, C. F., & Weiskrantz, L. Note on luminous flux discrimination in monkey and man. *Quarterly Journal of Experimental Psychology*, 1959, **11**, 49-53.

Gunter, R., Feigenson, L., & Blakeslee, P. Color vision in the cebus monkey. *Journal of Comparative & Physiological Psychology*, 1965, **60**, 107-113.

Hubel, D. H., & Wiesel, T. N. Receptive fields, binocular interaction and functional architecture in the cat's visual cortex. *Journal of Physiology*, 1962, **160**, 106-154.

Jacobs, G. H. Spectral sensitivity and color vision of the squirrel monkey. *Journal of Comparative & Physiological Psychology*, 1963, **56**, 616-621.

Johnson, H. M. Visual-pattern discrimination in the vertebrates. II. Comparative visual acuity in the dog, the monkey and the chick. *Journal of Animal Behavior*, 1914, **4**, 340-361.

Kinnaman, A. J. Mental life of two *Macacus rhesus* monkeys in captivity. I. *American Journal of Psychology*, 1902, **13**, 98-148.

Klüver, H. *Behavior mechanisms in monkeys.* Chicago: Univ. Chicago Press, 1933.

Klüver, H. Visual functions after removal of the occipital lobes. *Journal of Psychology*, 1941, **11**, 23-45.

Köhler, W. Optische Untersuchungen am Schimpansen und am Haushuhn. *Abhandlungen Preussische Akademie der Wissenschaften (Physikalisch-Mathematicsche Klasse)*, 1915, No. 3, 1-70.

Kohts, N. Recherches sur l'intelligence du chimpanzé par la méthode de 'choix d'après modèle'. *Journal de Psychologie Normale et Pathologique*, 1928, **25**, 255-275.

LeGrand, Y. *Form and space vision.* Bloomington: Indiana Univ. Press. 1967.

Locke, N. M. Perception and intelligence: Their phylogenetic relation. *Psychological Review*, 1938, **45**, 335-345.

Malmo, R. B., & Grether, W. F. Further evidence of red blindness (protanopia) in cebus monkeys. *Journal of Comparative & Physiological Psychology*, 1947, **40**, 143-147.

Medin, D, L., & Davis, R. T. Color discrimination by rhesus monkeys. *Psychonomic Science*, 1967, **7**, 33-34.

Miles, R. C. Color vision in the marmoset. *Journal of Comparative & Physiological Psychology*, 1958, **51**, 152-154. (a)

Miles, R. C. Color vision in the squirrel monkey. *Journal of Comparative & Physiological Psychology*, 1958, **51**, 328-331. (b)

Mishkin, M., & Weiskrantz, L. Effects of cortical lesions in monkeys on critical flicker frequency. *Journal of Comparative & Physiological Psychology*, 1959, **52**, 660-666.

Monjan, A. A. Chromatic adaptation in the macaque. *Journal of Comparative & Physiological Psychology*, 1966, **62**, 76-83.

Napier, J. R., & Napier, P. H. *A handbook of living primates.* London: Academic Press, 1967.

Ordy, J. M., Latanich, A., Samorajski, T., & Massopust, L. C. Visual acuity in newborn primate infants. *Proceedings, Society for Experimental Biology & Medicine*, 1964, **115**, 677-680.

Ordy, J. M., Massopust, L. C., & Wolin, J. R. Postnatal development of the retina, electroretinogram, and acuity in the rhesus monkey. *Experimental Neurology*, 1962, **5**, 364-382.

Ordy, J. M., & Samorajski, T. Visual acuity and ERG-CFF in relation to the morphologic organization of the retina among diurnal and nocturnal primates. *Vision Research*, 1968, **8**, 1205-1225.

Payton, C. R., & Blake, L. Difference limen for perception of the vertical in monkeys. *Perceptual & Motor Skills*, 1964, **19**, 455-461.

Pitt, F. G. H. The nature of trichromatic and dichromatic vision. *Proceedings of the Royal Society of London*, 1944, **1328**, 101-117.

Polson, M. C. *Spectral sensitivity and color vision in* Tupaia glis. (Doctoral dissertation, Indiana University, Bloomington, Ind.) Ann Arbor, Mich.: University Microfilms, 1968. No. 69-4797.

Polyak, S. L. *The retina.* Chicago: Univ. Chicago Press, 1946.

Pribram, K. H., Spinelli, D. N., & Kamback, M. C. Electrocortical correlates of stimulus, response and reinforcement. *Science*, 1967, **157**, 94-96.

Reid, J. B., Medin, D. L. & Davis, R. T. Perception of verticality by monkeys. *Journal of Comparative & Physiological Psychology*, 1965, **60**, 208-212.

Riesen, A, H., Ramsay, R. L., & Wilson, P. D. Development of visual acuity in rhesus monkeys deprived of patterned light during early infancy. *Psychonomic Science*, 1964, **1**, 33-34.

Rohen, J. W., & Castenholtz, A. Über die Zentralisation der Retina bei Primaten. *Folia Primatologica*, 1967, **5**, 92-147.

Rosenblum, L. A., & Cross, H. A. Performance of neonatal monkeys in the visual cliff situation. *American Journal of Psychology*, 1963, **76**, 318-320.

Schilder, P., Pasik, P., & Pasik, T. Total luminous flux: A possible response determinant for the normal monkey. *Science*, 1967, **158**, 806-809.

Schmidt, U. Untersuchungen zur visuellen Raumorientierung bei Totenkopfaffen *(Saimiri sciureus L.).* *Zeitschrift für vergleichende Physiologie*, 1968, **60**, 176-208.

Schrier, A. M., & Blough, D. S. Psychophysical studies of vision on monkeys. *In* D. E. Pickering (ed.), *Research with primates.* Beaverton, Ore.: Tektronix Foundation, 1963. Pp. 51-57.

Schrier, A. M., & Blough, D. S. Photopic spectral sensitivity of macaque monkeys. *Journal of Comparative & Physiological Psychology*, 1966, **62**, 457-458.

Shepard, W. T. Some mental processes of the rhesus monkey. *Psychological Monographs*, 1910, **12**, No. 5.

Shumake, S. A., Smith, J. C., & Taylor, H. L. Critical fusion frequency in rhesus monkeys. *Psychological Record*, 1968, **18**, 537-542.

Sidley, N. A., Sperling, H. G., Bedarf, E. W., & Hiss, R. H. Photopic spectral sensitivity in the monkey: Methods for determining, and initial results. *Science*, 1965, **150**, 1837-1839.

Sidley, N. A., & Sperling, H. G. Photopic spectral sensitivity in the rhesus monkey. *Journal of the Optical Society of America*, 1967, **57**, 816-818.

Silver, P. H. Spectral sensitivity of a trained bush baby. *Vision Research*, 1966, **6**, 153-162.

Spence, K. W. Visual acuity and its relation to brightness in chimpanzee and man. *Journal of Comparative Psychology,* 1934, **18,** 333-361.

Sperling, H. G., Sidley, N. A., Dockens, W. S., & Jolliffe, C. L. Increment-threshold spectral sensitivity of the rhesus monkey as a function of the spectral composition of the background field. *Journal of the Optical Society of America,* 1968, **58,** 263-268.

Stiles, W. S. Color vision: the approach through increment-threshold sensitivity. *Proceedings of the National Academy of Sciences,* 1959, **45,** 100-114.

Symmes, D. Self-determination of critical flicker frequencies in monkeys. *Science,* 1962, **136,** 714-715.

Symmes, D. Flicker discrimination by brain-damaged monkeys. *Journal of Comparative & Physiological Psychology,* 1965, **60,** 470-473.

Taylor, M. M. Visual discrimination and orientation. *Journal of the Optical Society of America,* 1963, **53,** 763-765.

Tigges, J. Sind alle Halbaffen farbenblind? *Naturwissenschaften,* 1961, **21,** 677.

Tigges, J. Untersuchungen über den Farbensinn von *Tupaia glis (Diard 1820). Zeitschrift für Morphologie und Anthropologie,* 1963, **53,** 109-123. (a)

Tigges, J. On color vision in gibbon and orangutan. *Folia Primatologica,* 1963, **1,** 188-198. (b)

Tigges, J. On visual learning capacity, retention, and memory in *Tupaia glis, Diard 1820. Folia Primatologica,* 1964, **2,** 232-245.

Treff, H. A. Tiefenscharfe und Sehscharfe beim Galago *(Galago senegalensis). Zeitschrift für vergleichende Physiologie,* 1967, **54,** 26-57.

Trendelenburg, W., & Schmidt, I. Untersuchungen über das Farbensystem der Affen. *Zeitschrift für vergleichende Physiologie,* 1930, **12,** 249-278.

Watson, J. B. Some experiments bearing upon color vision in monkeys. *Journal of Comparative Neurology & Psychology,* 1909, **19,** 1-28.

Weinstein, B., & Grether, W. F. A comparison of visual acuity in the rhesus monkey and man. *Journal of Comparative Psychology,* 1940, **30,** 187-195.

Weiskrantz, L., & Cowey, A. Striate cortex lesions and visual acuity of the rhesus monkey. *Journal of Comparative & Physiological Psychology,* 1963, **56,** 225-231.

Weiskrantz, L., & Cowey, A. Comparison of the effects of striate cortex and retinal lesions on visual acuity in the monkey. *Science,* 1967, **155,** 104-106.

Westheimer, G. Visual acuity. *Annual Review of Psychology,* 1965, **16,** 359-380.

Woodburne, L. S. Visual acuity of *"Saimiri sciureus." Psychonomic Science,* 1965, **3,** 307-308.

Yarczower, M., Wolbarsht, M. L., Galloway, W. D., Fligsten, K. E., & Malcolm, R. Visual acuity in a stumptail macaque. *Science,* 1966, **152,** 1392-1393.

Young, F. A. The effect of restricted visual space on the refractive error of the young monkey eye. *Investigative Ophthalmology,* 1963, **2,** 571-577.

Young, F. A., & Farrer, D. N. Refractive characteristics of chimpanzees. *American Journal of Optometry and Archives of American Academy of Optometry,* 1964, **41,** 81-91.

Zeigler, H. P., & Leibowitz, H. A methodological study of "shape constancy" in the rhesus monkey. *Journal of Comparative & Physiological Psychology,* 1958, **51,** 155-160.

Chapter 4

Hearing[1]

William C. Stebbins[2]

*Kresge Hearing Research Institute and
Departments of Otorhinolaryngology and
Psychology, University of Michigan,
Ann Arbor, Michigan*

I. INTRODUCTION

As recently as 1929, in an exhaustive reference work on the apes (Hominoidea), Yerkes and Yerkes could offer no quantitative evidence concerning the hearing of any member of this group of animals. Indeed, other than the fact that they could learn to respond to sound in the laboratory (Kalischer, 1912; Shepherd, 1910), there was no available information on hearing for any of the nonhuman primates in 1929. The Yerkes' remarks,

[1] Preparation of this chapter was supported in part by research grants (NS 05077 and NS 05785) from the National Institute of Neurological Diseases and Stroke.

[2] I am indebted to Drs. G. Gourevitch, B. Masterton, J. Vernon, and the editors for their useful comments on the manuscript, and to Medlars at the University of Michigan and the Primate Information Center at the University of Washington for their help in searching the literature.

although directed to the orangutan *(Pongo)* (Yerkes & Yerkes, 1929), are expressive of their general concern over the lack of information and, therefore, are relevant to this point:

> It is either to laugh or to weep if one contemplates this display of ignorance or of inaptitude and unpreparedness for skillful inquiry. Even as we present these exhibits we choose to laugh, for the situation strikes us as incredibly absurd. Nevertheless, instead of holding up to ridicule pioneers in anthropoid research, we naturally should prefer to say simply and solely, "Nothing worthy of mention in a scientific publication is known about receptivity and sensibility in the orangutan." This moreover is the fact we should like to leave vividly in the mind of every inquiring psychobiologist [p. 170].

Within 5 years, two psychobiologists at Yale, Elder (1934), a student of the Yerkes', and Wendt (1934), in the Laboratory of Neurophysiology, responded to the challenge and published the first systematic laboratory findings on hearing in nonhuman primates. Elder's subjects were three chimpanzees *(Pan troglodytes)*, and Wendt's included an olive baboon *(Papio anubis)*, a rhesus macaque *(Macaca mulatta)*, a sooty mangabey *(Cercocebus atys)*, and two spider monkeys *(Ateles paniscus)*. The work was exceedingly thorough, and the investigators were very conservative and cautious about their findings. This is not surprising, considering the somewhat primitive state of both acoustics and behavioral experimentation at that time. However, it speaks well for their thoroughness and rigor that their results are very close indeed to those of recent investigators who have all the advantages of improved technique in both the physical and behavioral sciences.

Though there may well be a yet undiscovered species exception, Elder's (1935) statement that the "difference in auditory sensitivity between man and other primates probably is a result of the latter's relative superiority in reception of high frequency sounds [p. 115]" clearly characterizes one major aspect of the hearing of nonhuman primates as we know it. This chapter is an account of the evidence that supports this statement, and is also a review of the experimental literature on hearing in nonhuman primates.

The experimental findings have been provided by only a few investigators. Several families of Prosimii (prosimians) and Anthropoidea (monkeys and apes) are represented. Despite widely varying behavioral procedures and different techniques for acoustic stimulation and measurement, rather clear and consistent results for the frequency range of hearing and the absolute sensitivity over that range have been obtained. There is little else concerning the hearing of nonhuman primates about which we can be as assertive. However, there is some interesting work beginning on other aspects of primate hearing. In addition, improved conditions in acoustic instrumentation and behavioral methodology have opened up a large number of hitherto untapped problem areas.

II. PROBLEMS OF MEASUREMENT AND ASSESSMENT

There are two primary concerns for the laboratory scientist interested in a behavioral analysis of sensory function. First, there are the procedural problems of instructing the subject, ordering the stimuli for presentation, and data treatment of the subject's responses. We need some degree of certainty that our measurements indicate something about the resolving power of the sensory system and are not confounded with motivational effects, instructional variables, or any of a large number of nonsensory influences. The second concern is with the instruments, particularly the acoustic transducers. It is always extremely difficult to evaluate the characteristics of the sound field in relation to the physical position of the subject in any experimental arrangement. Precise measurement of the stimuli in physical units appropriate to the source of energy is essential if we are to make any quantitative statements about the sensory acuity of an organism.

A. Animal Psychophysics

1. PRINCIPLES

The oldest and even now perhaps the most controversial area of inquiry in laboratory psychology, psychophysics, is heavily introspective and therefore firmly tied to language. The term "psychophysics" refers to a set of experimental data for humans on absolute and difference thresholds for various forms of energy and for the different sense modalities. It also clearly implies a set of experimental procedures by which these data are obtained. An integral part of these procedures is the verbal instructions that must be given to the subject prior to the experiment, and the rules governing the subject's response (usually verbal).

The use of language for instructing the subject, and for the subject's response, assumes an extensive and complex history of discrimination training with respect to language on the part of both experimenter and subject. The effect of language in psychophysical experiments has received comparatively little attention. The assumptions we make about the kind of control over behavior exercised by verbal instructions acting in a discriminative manner may not always be warranted. The role played by language and language behavior in the analysis of sensory function is far from clear, and herein lies one of the major enigmas for human psychophysics. Perhaps even more serious for the student of nonhuman sensory function is the fact that there are many sensory characteristics attributed to humans which cannot be defined independently of language. Loudness, as one of many examples, is characterized by a set of verbal instructions given to a trained human subject. For the linguistically underprivileged nonhuman primate, loudness defined in this manner cannot exist. In like manner, many of the so-called "psychological" attributes of sensation (brightness, pitch, volume, etc.) can have little species generality or biological significance in an evolutionary sense unless they can be separated from their purely linguistic frame of reference.

One of the advantages of constructing a psychophysics for lower animals is that one is freed from having to make any assumptions about linguistic repertoires, and except in generalization experiments, there is seldom need for concern about the prelaboratory discriminative history of the organism. The inability to rely on language creates an entirely new set of problems peculiar to animal psychophysics. Where no relevant discriminative history exists, one must be provided. The behavioral methodology (e.g., the conditioning procedures) for providing this history, and consequently producing the appropriate behavioral repertoire, becomes an important issue. It is necessary to find the correct substitute for verbal instructions. What are the training procedures necessary to bring the animal subject to the same point as his human counterpart following preexperimental instructions? It is these procedures that must transform the relatively untrained animal into a reliable psychophysical observer.

The experimental evidence for hearing in nonhuman primates, then, has depended upon procedures from two traditionally separate fields in psychology—one, human psychophysics; the other, learning and conditioning. The techniques of operant conditioning (Skinner, 1938) played a major role in the development of a nonverbal psychophysics, and these techniques have led to results that are, in most instances, as reliable as those obtained from trained human observers. For a discussion of the use of operant conditioning with nonhuman primates, see Kelleher's chapter (1965) in the first volume of this series.

The behavioral conditioning procedure should ensure first that the experimental animal is responding differentially in some way along the appropriate stimulus dimension. In the simplest example, the animal is trained to make one simple motor response in the presence of the stimulus (the "yes, I hear it, see it, etc." equivalent) and another response in its absence. Terrace (1966) has reviewed much of the literature on stimulus discrimination learning in animals. Many of the concepts and the methods now employed in animal psychophysics have come from early studies of discrimination learning (Keller & Schoenfeld, 1950, Chapter 5).

Over the years, there has been a certain amount of friction between students of animal behavior with those in the field claiming that insufficient attention has been given by those in the laboratory to the particular species characteristics of their subjects. The latter have, on occasion, countered by offering evidence that certain basic principles of behavior can be shown to apply over an extensive portion of the animal kingdom. Both views are appropriate here and considered in detail elsewhere (Stebbins, 1970a). It is unlikely that the basic conditioning methods or the psychophysical testing methods will need much alteration with different genera or species of primates. However, the living conditions, including diet, temperature, humidity, and housing, must be known and it is only in recent years that we have had sufficient information to keep some species alive in captivity. In addition to the basic requirements for healthy subjects, it is helpful to have some information about safety limits for food and water deprivation, effective reinforcers, and optimal experimental arrangements. For example, we have found water to be a useful, safe, and long-term reinforcer for tree shrews (Tupaia), as it is for the laboratory rat, but much less workable with the macaques whose behavior seems to be more efficiently maintained by food reinforcers such as nutritive pellets.

Few would deny the importance of knowing as much as possible about the behavioral and biological nature of their subject. Though the rhesus monkey has enjoyed the greatest popularity of any nonhuman primate as an experimental subject, there are not many who would recommend him highly where repeated close contact between subject and experimenter is an experimental requirement. There are other macaques who are more tolerant when, for example, earphones have to be carefully fitted on a daily basis. For the animal psychophysicist who is forever looking for ways to reduce variance in the behavior of his subject, accurate and detailed knowledge of their basic characteristics is indispensable.

2. Conditioning Methods

By examining the recent literature on the behavioral assessment of sensory function in animals beyond the limits set for the present chapter on one organism and one sensory modality, it is possible to get a wider perspective on the utility and effectiveness of various conditioning procedures. These procedures have been reviewed by Blough (1966), and more recently, and in greater detail, by Blough and Yager (1971) and Stebbins (1970a). Although it is true that many different procedures continue to be used, there are two that prevail in the current literature on animal psychophysics. Perhaps the most popular is the conditioned suppression procedure, which has been used successfully in studies of hearing in nonhuman primates (Masterton, Heffner, & Ravizza, 1969; Allen, Dalton, Henton, & Taylor, 1968) as well as in a variety of other experiments on sensory function in animals (Sidman et al., 1966; Hendricks, 1966; Ray, 1970; Smith & Tucker, 1969). Suppression or cessation of operant responding to a stimulus preceding unavoidable shock had originally been described in 1941 by Estes and Skinner, but only recently has the decrease in response rate been taken over as a useful reporting response in psychophysical experiments with animals. Different definitions of the change in response rate, which constitutes a reporting response, are used. The reliability of the procedure and the relatively good agreement with other findings serve as a sufficient basis to recommend it.

The second conditioning method uses a baseline maintained by positive reinforcement, and a clearly-defined observing response. The paradigm is a two-link behavioral chain in which the first response is followed on some schedule by stimulation and the second response (the report) leads to reinforcement. The method probably owes much of its success to the use of an observing response, which usually serves to place the animal in the best position to receive stimulation and to react to it quickly. Gourevitch (1970) and Stebbins et al. (1966) have found the procedure particularly useful in work on hearing in nonhuman primates; others have used it to examine visual and auditory function in other animals (Dalland, 1965; Gourevitch & Hack, 1966; Berkley, 1970; Nevin, 1964; Yager, 1968).

Avoidance procedures, in which the stimulus reporting response serves to avoid shock, have been described for auditory threshold determination in monkeys (Behar, Cronholm, & Loeb, 1965; Fujita & Elliott, 1965; Martin, Romba, & Gates, 1962; Clack & Herman, 1963) and in other sensory experimentation (Krasnegor & Brady, 1968; Hanson, 1966).

Perhaps because there is a better understanding of basic conditioning principles and methods for obtaining effective stimulus control with food reinforcement (Terrace, 1966) than with shock, there appears to be a clear preference by most investigators for baselines maintained by the former. However, I have been unable to find any very clear evidence in the literature of animal psychophysics to support the use of one type of reinforcement over another. In our laboratory, we have obtained comparable auditory thresholds with food-reinforced responding and shock avoidance as have Fujita and Elliott (1965). Early proponents of the use of shock (e.g., Harris, 1943) have claimed more rapid results without the motivational changes associated with the positive reinforcers such as food and water. On the other hand, some prefer to avoid the emotional behavior produced by shock and also the potential hazard involved in using shock with subliminal stimulation. To the subject, these are shocks uncorrelated with specific stimulus events and the effect has sometimes been a complete generalized suppression of all operant behavior. It has been our limited experience that thresholds obtained under food reinforcement baselines are somewhat more stable over time than those obtained under a shock avoidance regimen. I hasten to point out that the use of conditioned suppression constitutes a third procedure which, although it obviously contains elements of both avoidance and food-reinforced responding, must be considered separately. The use of shock in the suppression procedure is independent of the organism's behavior, whereas in the shock-avoidance procedure, the behavior is maintained by its effect in terminating the signal which has preceded the shock.

Recently, Dalton (1967, 1968) has compared results obtained with the conditioned suppression technique to those obtained with both a galvanic skin response (GSR) and an averaged evoked cortical response (EEG). Training time was a factor, and additional time was required for the conditioned suppression procedure before thresholds were obtained. However, Dalton reports considerably greater sensitivity and replicability for the conditioned suppression procedure. Semenoff and Young (1964) have described the use of the GSR conditioned to pure-tone stimulation as a method for threshold determination in the monkey. Their findings for the macaque are not in good agreement with those of other investigators. Primarily because of savings in time, EEG and GSR audiometry would appear to be the panacea for threshold testing in animals. However, questions concerning the reliability of the techniques, the presence of artifacts, and the interpretation of the results preclude serious consideration of these approaches at this time.

An important feature of most of the present techniques is their need for the subject to be in a relatively fixed position with respect to the stimulus source. Consequently, the stimulus is more precisely located with reference to the sensory receptors and more exact stimulus measurement is possible. By contrast, in some of the earlier procedures involving use of, for example, the shuttle or double-grill box, considerable spatial movement was required on a trial-by-trial basis. The careful utilization of more recent free-operant conditioning procedures has also permitted almost continual observing by the subject with a gain in reliability and in the amount of information acquired per unit of time.

3. PSYCHOPHYSICAL METHODS

Once the discriminative behavior of the animal has stabilized, the procedures that are used for varying the stimulus along some physical dimension and the subsequent treatment of the behavioral data are more or less the same as in human psychophysics.

The purpose of the behavioral conditioning procedures is first to develop stimulus control (e.g., discriminative responding), and then to maintain this control during threshold determinations, while some characteristic of the stimulus is being altered by the selected psychophysical procedure. The probability of a breakdown in the discrimination is enhanced as the stimulus energy is decreased, or as two stimuli on the same physical continuum are brought closer together for the first time. It sometimes becomes necessary to modify the standard psychophysical procedure to avoid complete loss of stimulus control. Clearly, extreme care is necessary in order that the change does not bias the procedure in some way that would be reflected in the sensory data. A good example of the kind of problem that can arise is seen in the use of the method of limits. Frequently, the ascending series of stimulus presentation will be omitted, since it necessitates presenting the stimulus presumably below the animal's threshold on several consecutive trials, making it difficult to maintain good stimulus control. However, evidence from human subjects indicates that the use of the descending method alone leads to low estimates of the threshold. Perhaps in this situation, the method of constant stimuli would have been a better choice. In other circumstances, it may be possible to compromise to the extent to which stimulus control is maintained, and yet not influence the basic sensory data by the change in psychophysical procedure.

Another way in which bias may be introduced is peculiar to animal psychophysics, and may be more of a problem with the nonhuman primates because of their many known and many more assumed similarities to man. The obvious trap, and yet one which is not always easy to detect, is that the experimental design for the animals may in some way be influenced by such assumed similarities to man or even to some other closely related species. It may be a great temptation to preguess the experimental results, and suggest that the organism's sensory acuity is like that of man, for example. Such bias could be easily transmitted to the experimental animal, which in turn produces a humanlike acuity function and reinforces the experimenter's unwarranted hypothesis. The problems attendant upon starting with a new and untested species are considerable. The initial choice of stimulus values in such a situation can be a difficult one to make.

Many of the problems of human psychophysics are simply transferred to animal psychophysics. We have recent laboratory evidence, however, that variables such as psychophysical method used, step size for stimulus variation, type and schedule of payoff, or reinforcement have not affected thresholds obtained in studies of nonhuman primate hearing (Stebbins, 1970b). In fact, the variability in the sound field (due to small fluctuations in the speaker's positional relation to the subject's ear) and stimulus measurement error, particularly at the higher frequencies of stimulation, may have a greater effect on the data than any of these procedural variables.

The definition of threshold is traditionally statistical in that it represents an averaged estimate of the subject's responses over a large number of stimulus presentations. Deciding which value of the stimulus will be considered as threshold is, to a certain extent, arbitrary; the decision depends on the testing method used. An extensive literature exists on the subject (see Woodworth & Schlosberg, 1954). In using the standard human psychophysical procedures, it seems most appropriate to adopt the standard criteria: the 50% point from the psychophysical function for the constant stimulus method, and the averaged transition point between a correct detection and a miss for the tracking method and the method of limits.

Selection of a psychophysical method for use in the determination of sensory acuity in animals depends on the conditioning method that the experimenter has adopted, and, to a certain extent, on whether he demands speed or accuracy.

Most of the current work utilizes a conditioned somatic or operant response, and the requirements are relatively similar to those for a human verbal response. As with humans, if the emphasis is on accuracy and reproducibility, the method of constant stimuli is chosen. However, it is time consuming and provides a great deal of wasted information. The tracking method, in which each positive response serves to decrease stimulus intensity, and each negative response increases it, is unquestionably faster. With this method, an entire sensitivity function can be obtained in a single session. Variability, however, is relatively high under this procedure because it keeps the stimulus close to threshold, continually forcing the subject to make an extensive series of very difficult discriminations. It is safe to say that one of these two methods, or some slight modification that would lessen their disadvantages, can be used most effectively in working with animals.

New problems may arise when other than operant or instrumental conditioning methods are used. If an analog signal recorded from the subject (GSR, EEG, or even EKG) is accepted as an indicator response, the problem of defining threshold becomes substantial. It is no longer possible to start with a simple binary event (the occurrence or nonoccurrence of a lever response by the subject); decisions must be made about the characteristics of the analog response. Usually, threshold is defined in terms of amplitude of the waveform (Semenoff & Young, 1964), or even as the experimenter's visual discrimination threshold for detecting the signal above the baseline noise level on an oscilloscope screen (Dalton, 1968). Perhaps due to these difficulties, it is not surprising that the results obtained with the GSR or the EEG are not in good agreement with other findings on the auditory sensitivity of nonhuman primates.

B. Instrumentation

To a certain extent, the presentation and measurement of acoustic energy is a problem in physical acoustics (Beranek, 1949; Hirsh, 1966; Keast, 1967). The use of nonhuman primates as experimental subjects creates some interesting complications that are worthy of consideration.

Precise measurement of the sound pressure at the ear while it is being stimulated is a seldom-achieved ideal. Sound may be presented from a speaker at a distance from the subject and propagated openly through the air, or in a closed system (by tube from a speaker), or by earphones mounted on the head. For accurate specification of the sound entering the external ear, the earphone is preferred. Earphones of a size to fit the smaller nonhuman primates can cover only a small part of the frequency range of these animals. A primate weighing 5 lb or more can be fitted with standard size human earphones. There are some commercially available that have a sufficient frequency response, albeit not flat, to cover the nonhuman primate's extended frequency range of hearing. Precise calibration of the output of these earphones in position over the ear canals, especially at the higher frequencies, is quite another problem.

In fact, the most serious difficulties for measurement exist at the higher frequencies. At the shorter wavelengths, the sound is highly directional, and small movements of the animal's head can produce extremely large fluctuations in sound pressure at the ear. The use of a wide-range speaker, free field, is subject to this kind of error if the animal is permitted any movement, including that of the pinna. The use of a conical tube to guide the sound is seldom practical in behavioral experiments, since it also demands an immobile subject. In addition, the tube is subject to standing waves or various interference patterns produced by reflection from the tube walls, which make stimulus specification unreliable. Earphones, since they move with the subject, are advantageous only if they can be maintained in proper position over the entrance to the ear canals for the course of the experiment. At the higher frequencies, standing waves are also present in the ear canals.

Figure 1 illustrates our experimental arrangement for the macaque. The earphones are mounted on special holders with universal joints. The monkey is permitted some head movement in all three dimensions, and the earphones move with the head. His nose is positioned between two vertical bars to the front and his head is held in the back by a contoured plastic support. The animals adapt to this degree of restraint with little difficulty, and they will endure it for the duration of the experimental session (about 2 hr). Figure 2 is a schematic diagram of the sound measurement system. Briefly, a small probe tube connected to a condenser microphone is inserted through the earphone cushion and positioned in the earphone center and at the opening of the external canal. Thus the actual measurement is performed with the phones on the animal (however, not during the experimental session, and with the aid of a tranquilizer). In this manner, characteristics of the monkey's external canal and the effect that this tubelike structure has on the entering sound wave are taken into consideration. For pure tones, the use of a wave analyzer is important—for measurement of the energy contained in the tone itself, and also for specification of its harmonic structure. The importance of measuring the harmonic content of the test tone cannot be overemphasized. Should the animal be 40 dB less sensitive at 250 Hz than at 1 kHz, and, if the third harmonic (1 kHz) for the earphone is found to be only 30 dB below the fundamental (250 Hz), then the threshold measurements presumably taken at 250 Hz, in fact, reflect the animal's sensitivity at 1 kHz. The problem is discussed by Dalland (1965), who has encountered it in his work with the bat. A sound-level meter (often used) is much less desirable because it is not frequency selective, and will simply totalize all of the acoustic energy incident on its microphone at a given time. A frequency counter and voltmeter continuously monitor the frequency and amplitude (root-mean-square volts) of the tone into the earphones. If an

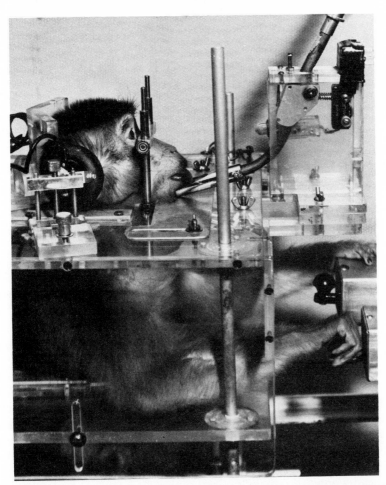

FIG. 1. The subject seated in a restraining chair during threshold determination. Earphones, response keys, and feeder tube are shown. The monkey has just reported the tone and has extended his tongue to receive food reinforcement. [From Stebbins *et al.*, 1969].

oscilloscope is used the amplitude measurements are peak-to-peak or 1.4 times the root-mean-square value. Periodic calibrations of the phones as described above give the actual sound pressure incident at the ear for a given frequency and voltage across the phones.

Where a large speaker is used for free-field presentation, the condenser microphone is positioned in place of the animal in the approximate location of the head or ears. In situations where the animal is unrestrained, measurements are taken with the microphone at several points in the sound field and the mean

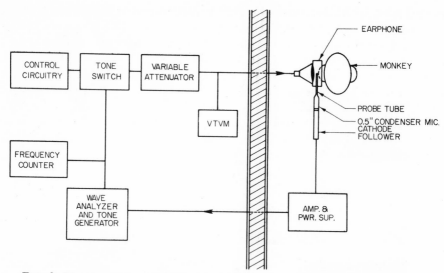

FIG. 2. Diagram of experimental arrangement for stimulus control and measurement. [From Stebbins *et al*, 1969].

and range of the measurements in decibels relative to a standard reference pressure are presented.

The references cited at the beginning of this section provide a more detailed description of some of the finer points of acoustic measurement. Unfortunately, none of these references has specific relevance to studies of hearing in nonhuman primates. Perhaps Hirsch (1966) comes the closest in his discussion of acoustic measurement for research with humans. Presently, the only alternative is a review of the methods sections of the various experiments cited in the present chapter. However, the condensed nature of these descriptions makes this an unsatisfactory option.

Clearly, one of the obstacles to auditory research with nonhuman primates is the difficulty in presenting to the subject and measuring (with sufficient precision) the acoustic stimulus. In all fairness, a share of the blame for this state of affairs is due to an inadequately developed technology. Even the best instrumentation is often not satisfactory. Specifications for characteristics such as harmonic distortion, frequency drift, signal-to-noise ratio, and frequency range are not yet high enough. Particularly important for animal work will be effective miniaturization of the acoustic transducers without loss of fidelity. Earphones that are inserted into the external canal are still too large for most animals; in addition, such instruments have a very limited frequency response.

It may well be that the difficulties involved in both accurate and reliable specification of the acoustic stimulus and in behavioral training and stimulus programming have been overstated here. It is true that most of the data, at least for absolute intensity thresholds, are in good agreement, and this promotes additional confidence in the procedures. However, while there continue to be questions concerning measurement accuracy, there must remain some reservations about even the most convincing evidence concerning hearing in nonhuman primates. It is no longer acceptable to report sound intensity in volts across a speaker or to use the experimenter's ear as a reference. With reasonably good instrumentation available (particularly for the sonic frequencies, that is, those below 20 kHz), it would be inexcusable to settle for less. Such equipment is sufficiently common that it can be borrowed if not purchased.

III. HISTORICAL EVIDENCE AND FIELD OBSERVATIONS

A. Early Conceptions and Speculations

Early primatology in the 19th and early 20th century, like most of biological science, was very much under the influence of Darwin. Collection of specimens, anatomical description, and taxonomic assignment were the order of the day. Behavior was either described in rich literary fashion from anecdotal evidence, or inferred from anatomical structure.

As early as 1831, Warwick (1832) had an opportunity to observe a chimpanzee and an orangutan for a prolonged period in captivity. "The hearing of both animals was remarkably acute, catching the most indistinct noise at a considerable distance; and their knowledge of sounds was accurately shown; as, on hearing the footsteps, on the stairs, of persons with whom they were acquainted, they ran towards the door before it was opened [p. 308]."

Wallis (1897) presented a series of anatomical sketches of the external ear of man and the higher primates. In dramatic metaphor, he likened the primate external ear to an outpost; once important, but no longer essential, from which the garrison is withdrawing. With the notion that structures undergo progressive degeneration as the result of loss of function, and the additional assumption that size of the external ear was somehow related to hearing acuity, some inferences about primate hearing began to appear. Pocock (1905) and later Sonntag (1924) affirmed the relation between external ear size and habits. They suggested that the largest ears go to the chimpanzee and gibbon, for they require high auditory acuity to escape from their enemies on the ground. The orangutan is arboreal and therefore functions well with almost no visible signs of an external ear. Though the gorilla may appear the exception, being terrestrial and possessed of small ears, it was pointed out by Pocock and Sonntag that here is a formidable animal who can easily defend itself, and therefore to whom acute hearing is of little consequence.

B. Field Observations

Recent research on nonhuman primates outside the laboratory is largely devoted to group study of these animals, either in the field or in some form of compound with as many field characteristics present as possible. Much of the work has concentrated upon the many interactional and organizational systems of the social group. One system that is receiving an increasing amount of attention is communication (Marler, 1965). With the recent technological advances in electronics and acoustics, particularly in the area of sound recording and measurement, there has been a predictable move to studies of auditory communication in nonhuman primates. As Marler has pointed out, auditory signals are now more easily described and quantified, and more accurately preserved than, for example, visual or olfactory signals. Most interesting and suggestive are some recent recordings of primate calls by Struhsaker (1967, 1970). He has reported harmonics as high as 23 kHz in the call of the vervet monkey *(Cercopithecus aethiops),* and a frequency range extending to 32 kHz for the chip call of the dwarf galago *(Galago demidovii).* Although there is no evidence that these primates make use of anything other than the fundamental frequency of a call, nevertheless, the presence of high-frequency harmonics suggests at least the possibility of their significance in communication. Struhsaker has described the conditions under which some of these high-frequency calls occur. For example, the proximity of a predator, or the spatial separation of an infant from his mother, may evoke a call with a fundamental frequency as high as 16 kHz. It is interesting to consider that in heavily foliated environments, pulsatile high-frequency sounds may be very difficult to localize. The shorter wavelengths are more easily broken up by the limbs, leaves, and branches. Perhaps the caller is thus less vulnerable to the nearby predator.

Unfortunately, despite the concern with the intricacies of the acoustic signals emitted by primates and their discriminative function with respect to the behavior of the receiver, there seems to be a singular lack of attention paid to the receptive capacity of the hearer. Obviously it is not feasible to make precise audiometric measurements in the field. However, at least some form of more rigorous assessment might be possible along with the detailed and complex kinds of observations that are carried out on other aspects of primate behavior in the field. There seems to be the implicit assumption, at least for the study of vocal communication, that the hearing capabilities of nonhuman primates can be considered identical to those of man. The assumption may have been based on little more than the observation, and subsequent deduction, that since man can hear and attend to the signals transmitted by the lower primates *ipso facto,* they should hear as he does.

To find even the most speculative kind of comment about primate hearing in the field literature is difficult. Donisthorpe (1958) was surprised at how close she was able to approach gorillas without disturbing them. Schaller (1963) disagrees with Donisthorpe's rather poor opinion of the gorilla's acuity, arguing, on the basis of several observations of the distance at which a gorilla group can be disturbed by strange, often manmade sounds, that the auditory sensitivity of the gorilla is comparable to man.

On the basis of laboratory work, it is now well established that the nonhuman primates have a more extended frequency range of hearing than man. It would be of great interest to find out if this extended capacity has any significance for the primate in his natural habitat. In all fairness to the field investigator, the problem of free-field measurement of frequencies in the ultrasonic range (above human limits) is extremely difficult and perhaps even impossible at the present time. Nevertheless, even reports of high-frequency communication among primates, such as Struhsaker's, or of an obvious response to high-frequency stimulation, would be of value.

IV. LABORATORY EVIDENCE

A. Range of Hearing; Absolute Threshold

In 1933, on the basis of a now classic experiment, normative data on the nature of the human audibility function were published by two Bell Laboratory scientists, Sivian and White (1933). The absolute intensity thresholds, which they obtained at frequencies from 60 Hz to 15 kHz, under excellent laboratory conditions, are very close to the currently accepted international standards (ISO) for human hearing. The data can be seen in Fig. 6 and will be described more fully later under species comparisons. At almost the same time, Elder (1933, 1934) presented comparable data for the chimpanzee, as did Wendt (1934) for several species of Old World monkeys.

There were few laboratories in the world at that time that had the equipment or the technical skills necessary to make acoustic measurements as precise as those of Sivian and White. Both Elder and Wendt, in the only option available to them, used man as a standard or reference for their sound measurements, as will be described later.

There is didactic value in examining the historical development of experimental methodology. Unquestionably, much of the improvement in the methods of sensory experimentation can be traced to technological progress. However, in sensory work with animals, many procedural changes are not merely related to the possession of better instruments. Rather, such changes can be traced to a continual shaping process that takes place between the experimenter and the behavior of his subject. Procedures that produce stable sensory data quickly, that yield minimal behavioral variance, and that lead to the experimenter's ability to support the validity of his results can be expected to have a fair chance of survival in the laboratory.

In the published work of Elder and Wendt, there is a dominant concern with methodology, and one can readily see the first stages in the development of behavioral procedures for measuring hearing of nonhuman primates. It becomes apparent very quickly from their description that some methods are effective and others must be modified on the basis of the subject's behavior. The difficulties they encountered, the solutions they proposed, and some of the insights they had justify a fairly extensive treatment of their work in this chapter. For anyone rash enough to consider working in this area, there is much to be learned from a more complete and first-hand examination of their articles.

Wendt selected human subjects and obtained thresholds from them under free-field conditions with the same experimental arrangement he had used for his monkeys. The relative values for sound pressure (readings from an attenuator) thus obtained for the human subjects were transformed into absolute values by reference to the Sivian and White function for free-field testing (binaural M.A.F. for random horizontal incidence, Sivian and White, 1933, p. 313). Finally, the relative threshold values for sound pressure for the monkeys were assigned absolute values on the basis of their deviations from the values for the human subjects (Wendt, 1934, p. 34). Wendt (1934), of course, admits to the assumption that his human subjects "had the same average limens as those quoted by Sivian and White [p. 33]."

Elder used one of the earliest audiometers for humans, with human earphones fitted to his chimpanzees. The threshold functions for his animals were then compared to average human threshold [as reported by Fletcher (1929) and obtained with the same audiometer and earphones]. In fact, these values are strikingly close to those found by Sivian and White with earphones (their monaural M.A.P. function, 1933, p. 313).

We have transformed Elder's relative intensity values into absolute values on the basis of the sound pressure measurements that he reports for the audiometer. These data, together with Wendt's[3] and some of our own more recent data (averaged for four macaques from Stebbins, Green, & Miller, 1966), are presented in Fig. 3. Our results were obtained, and sound pressure measurements made, with all of the advantages of modern acoustic instrumentation and sound-treated environments.

It is interesting to note that despite extreme differences in instrumentation and behavioral technique, and in consideration of the fact that neither Wendt nor Elder measured their sound source directly in relation to their animals, the three audibility functions are strikingly similar. Both chimpanzees and Old

[3] I have made a slight change in Wendt's data by using the Sivian and White function for $0°$ azimuth rather than the one for horizontal incidence which he uses. This affects primarily the threshold value at 8 kHz.

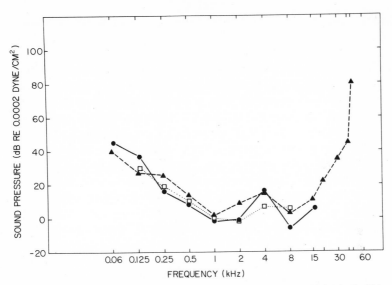

FIG. 3. Auditory threshold functions for the monkey [Wendt, 1934 (●); Stebbins *et al.*, 1966 (▲)] and the chimpanzee [Elder, 1934 (□)].

World monkeys show maximum auditory acuity in the 1- to 8-kHz range with a loss (or "notch") of as much as 15 dB at about 4 kHz. The agreement between these three functions is as good as that between the four animals of Stebbins *et al.*, whose data were averaged for the figure.

Elder estimated that his chimpanzees were between 3 and 7 years of age. Training and testing methods, to some extent, were developed on an empirical basis during the course of experimentation and tailored to the individual animal. The obvious methodological criteria were maintenance of responding and good stimulus control. During threshold testing, it was not uncommon for the animal to stop responding or for "unpredictable variations in sensitivity [1934, p. 167]" to occur. Though perhaps less severe, these behavioral changes are not unlike those Pavlov (1927) described for his dogs under discriminative training conditions, which led to "experimental neurosis." The chimpanzees refused to work and exhibited behavior similar to that which occurs in children's temper tantrums.

Subjects learned to respond to a pure tone, presented through an earphone, by pressing a key. A trial was signalled by uncovering the key and making it accessible to the subject. The presentation probability of a tone on any trial was 50%. A response to the tone was followed by food; responses in the absence of tone were simply not reinforced.

Elder (1934) regarded the use of a "ready," or trial signal, as the most important part of this procedure. He felt that the degree of stimulus control necessary to maintain accurate responding to signals near threshold could not be maintained over long periods of time. The use of this visual "ready signal" seemed "particularly advantageous with the chimpanzee subjects because it allowed them periods of relaxation [p. 165]." Elder notes that under these conditions the attention of the animals was excellent. The important clue to the success of this procedure appears later in his description of method—"the time interval between trials was not constant but was determined by the animal's readiness to respond [p. 166]." Apparently, the initiation of a trial by uncovering the key was contingent upon some variant of the response of "holding still" on the part of the animal. Thus, conditions for best listening were assured. This particular contingency would seem to be a very effective one for ensuring low thresholds and for reducing response variability; however, it must have produced a considerable strain on the experimenter's patience. There are a number of devices currently available, such as strain gauges, and photoelectric cells, which would permit automation of this "holding" response. Movement with respect to the stimulus during a trial is a problem that does not receive enough attention from investigators working on sensory problems with animals.

In Elder's experiments, thresholds were obtained by an early form of the tracking method—the method of serial groups. Stimulus intensity remained the same for ten consecutive trials. The intensity was decreased if the animal reported correctly on more than five trials out of ten, and increased if less than five trials were responded to correctly. Threshold was defined as the stimulus value to which the subjects responded correctly on 50% of the trials. Elder was quick to point out that considerable behavioral variability during the experiment precluded strict adherence to one testing procedure. He was also unwilling to be held to the somewhat inflexible dictates of the human psychophysics of the day. "The effect of minor deviations from 'orthodox psychophysical procedure' is a matter which does not concern us greatly. The problem of getting the subjects to work most efficiently is more important [1934, p. 176]."

Elder does suggest that if there is error in his data, it is in the direction of low threshold estimates; thus his results represent "maximum sensory capacity" for the chimpanzee. One can readily agree with his statements, "The regularity and consistency of the responses, relative absence of false responses, and the degree of precision of the psychometric functions are dependable criteria," and "In these respects some of the data compare quite favorably with that of a good human subject in the same situation [1934, p. 176]."

Elder's subsequent concern was with the most obvious difference he (and also Wendt) had found between man and the nonhuman subjects—the apparent superiority of the latter in acoustic sensitivity to high-frequency stimulation. In a paper published in 1935, Elder attempted a comparison of the upper frequency limit of hearing for man and chimpanzee. His subjects were three children between the ages of 5 and 12, and two of the chimpanzees he had used in his previous experiment. Although tonal frequency was measured, no measurements of sound pressure level were made. The testing method was essentially the same as described previously. For the children, the upper-frequency limit of hearing ranged between 22.6 and 23.7 kHz, and for the chimpanzees, it was between 26 and 33 kHz. With clear separation between his two groups, Elder could at least suggest that man could be distinguished from his nearest primate relative by a somewhat more restricted hearing range due to relative insensitivity to frequencies above 20 kHz.

Wendt's results (1934) on three species of Old World monkeys (a rhesus monkey, an olive baboon, and a sooty mangabey) confirmed and added support to Elder's findings. The similarities and differences between man and lower primates (particularly the high-frequency superiority of nonhuman primates) were verified. However, it was not possible on the basis of Elder's and Wendt's data to make any distinctions among the nonhuman primates themselves—prosimian, monkey, or ape. With respect to their auditory capacity, they had to be grouped as one class in contradistinction to man.

Wendt's published report of his work can be considered one of the truly epic reports in the literature of early experimental psychology. His procedure is sufficiently detailed to allow almost exact replication. Scale drawings from different views of the entire experimental arrangement are included. Sample protocols and tables and graphs of individual data are presented. Every aspect of the equipment, the training, and of his extensive procedures is meticulously described.

Wendt's methods were similar in some respects to Elder's. The raising of a curtain acted as a "ready" signal and was followed at an interval of either 8 or 16 sec by a tone of 4-sec duration. A drawer-opening response during the tone produced food; a premature response to the drawer terminated the trial and postponed the next one for 1 min. A method of limits (descending series only) was used; the tone was decreased in 2-dB steps until the animal failed to respond and then increased by 10 dB for another series. Average thresholds were calculated on the basis of the lowest tone-intensity in each series to which the animal reacted.

Two of the most interesting features of Wendt's procedure were the use of a "ready" signal and the punishment of incorrect "anticipatory" responses by delay of the next trial. Wendt, like Elder, never employed shock and, in fact, was strongly opposed to its use: "It apparently results in far different behavior consequences than the experimenter intends [1934, p. 23]." Although Wendt does report a fairly high incidence of false responses, the extinction and trial delay procedures seemed to have the effect of reducing these responses during continued training and testing. Wendt presents a very thorough analysis of the problems he encountered, and their causes (1934, pp. 36–45).

In the discussion of his data, Wendt notes the obvious similarities to Elder's audiometric measurements on chimpanzees. He calls attention to the discontinuity or loss in sensitivity at 4096 Hz (the "4096 dip"), which was also evident in Elder's data, but is not seen in human data. He suggests further study of this particular phenomenon. We find this discontinuity in our animals, and it may be as high as 20 dB; we also see a lesser break at 2 kHz in some animals. The break is evident in individual human data (Lawrence & Yantis, 1956; Ross, 1967), but is of lesser magnitude and is lost in averaging over many subjects, since it may not always occur at the same frequency. Parenthetically, a "notch" or "dip" is evident in the audiogram of many different animals, but it appears at different frequencies. In the cat, it was once thought to reflect the resonant frequency of the cat's bulla. This explanation has lost some credence, since primates without a bulla show the same phenomenon.

In the only other experiment in the literature on the hearing of nonhuman primates between 1934 and the 1960's, Wendt's data for the monkey found support from Harris (1943). Harris' experimental design differed greatly from either Elder's or Wendt's. He essentially adopted Culler's "motor conditioning method" (Culler, Finch, Girden, & Brogden, 1935), which was an early variant of a discriminative avoidance procedure.

The subjects, macaques, were placed in a suspended cage designed as a stabilimeter. Although the definition of a response is not clear, apparently any behavior of the animal which caused visible motion of the cage during tonal stimulation was considered sufficient to avoid shock. Threshold testing was done using the method of serial groups described earlier in this section. The serious problem of punishing the animal for failure to respond to subliminal stimuli was somewhat lessened by only occasionally shocking at low stimulus values. The data are in accord with those of Elder and Wendt with the exception of the point at 8 kHz, the highest frequency he tested, where the threshold is 15–20 dB lower than that found by Elder or Wendt. Further, the failure to replicate this low threshold value (about 20 dB below 0.0002 microbar) by any of several recent investigators (Behar *et al.*, 1965; Clark & Herman, 1963; Fujita & Elliott, 1965; Semenoff & Young, 1964; Seiden, 1958; Stebbins *et al.*, 1966) suggests the possibility of error in the measurements taken at 8 kHz. Tones were presented free field from an overhead speaker, sound power measurements with a sound-level meter were made in various parts of the cage, and average values were used in the final calculations. Harris, on the other hand, argued that his animals were younger and therefore may have had more acute hearing.

Harris noted that his training and testing procedures were considerably more rapid than Wendt's or Elder's and avoided "motivational disturbances." However, he does not describe false responses, and, if they did occur, it is not clear how they were treated. Perhaps a more serious problem is the fact that the definition of a response depends on the judgment of the experimenter at the time of the trial. Harris (1943) does say that "the true conditioned response is usually unmistakable in pattern even to persons unfamiliar with the behavior of monkeys in this and other situations [p. 261]." Under these conditions, replication in another laboratory is not always a simple matter. To my knowledge, this particular procedure has not been used again for audiometric testing of monkeys.

In an unpublished doctoral thesis, Seiden (1958) described his work on the auditory sensitivity of the marmoset (Callithrix jacchus). His study represented the first attempt to cover the entire frequency range of hearing for any species of nonhuman primate. Seiden's data are the first to indicate clearly the extent of the nonhuman primates' acoustic superiority over man at higher tonal frequencies. The marmosets' range extends to about 40 kHz. Where the frequencies studied overlap those used in the previous monkey studies, the agreement in the results is good. The "dip" or decrement in acuity near 4 kHz is evident, as is the superior sensitivity to man at 8 kHz and above.

Sound stimuli in Seiden's experiments were presented free-field under excellent laboratory conditions. Stimulus control was obtained with a shock-avoidance procedure in a double-grill box with the speaker located over the center of the box. Sound measurements were carried out with a condenser microphone and wave analyzer, and voltage readings (similar to those by Harris) taken from several positions in the cage were averaged. Like his predecessors, Seiden delayed the onset of a trial until his animals were relatively inactive. The shock-avoidance response consisted simply of crossing from one side of the cage to the other when a tone was presented. Interestingly enough, the use of shock was discontinued following avoidance training, and threshold testing by the "up and down" or tracking method was carried out in extinction. Statistical analysis revealed no systematic change in behavioral performance over the course of testing that would suggest that the avoidance response had not yet begun to show the effects of extinction.

In the decade following Seiden's thesis, there have been at least eight studies dealing with some aspect of the auditory acuity of nonhuman primates. Several have been primarily concerned with the use of a new conditioning or testing procedure, and have presented only limited data on auditory acuity (Clack & Herman, 1963; Allen et al., 1968; Martin et al., 1962; Semenoff & Young, 1964). These were mentioned earlier under methodological considerations. Several investigators have produced more extensive acuity data for several species of Old and New World monkeys (Behar et al., 1965; Fujita & Elliott, 1965; Stebbins et al., 1966); and Heffner, Ravizza, & Masterton (1969a, b), with a distinctly different procedure, have provided us with the first findings on the hearing of a number of prosimians.

Behar et al. (1965), using a shock avoidance procedure developed by Clack and Herman (1963), described threshold functions for rhesus monkeys from 50 Hz to 31.5 kHz. However, it is clear from their data that the upper-frequency limit for their subjects is considerably higher. Stimulation was free-field, and precise measurements of sound pressure were made at the various ear positions for the monkey. Instead of averaging measurements from the different locations, the figure for the highest energy level was adopted, thus producing the most conservative threshold estimates. A single lever tracking procedure was used, and shock for failure to avoid was programmed only 30% of the time. Again, the results are close to those of previous studies. The function is somewhat more flat in the midrange frequencies without the pronounced contrast between the "dip" at 4 kHz and extreme sensitivity at 8 kHz.

Additional confirmation for the monkeys' auditory sensitivity function and high-frequency receptivity came the same year from an extensive study by Fujita and Elliott (1965). Their subjects were rhesus and cynomolgus *(Macaca fascicularis)* macaques and squirrel monkeys *(Saimiri sciureus)*. Thresholds between 62.5 Hz and 32 kHz were obtained with a variety of procedures, and again it was evident from the data that all the animals would respond to still higher frequencies. Fujita and Elliott compared three different behavioral training methods. In one, several squirrel monkeys were trained in shock avoidance in a double-grill box with a procedure similar to that reported by Seiden. In a second procedure, macaques and squirrel monkeys were conditioned with a single-lever avoidance method. Tonal stimuli were presented for 5 sec and followed with shock; a lever response to the tone avoided shock and terminated the tone. Shock was administered for responses between trials. The third procedure employed all three species and used a single lever to report tonal stimulation. Correct reports were followed by food reinforcement; failure to respond simply ended the trial without reinforcement. Intertrial responses delayed the subsequent trial. For hearing testing, stimulus intensity was decreased until the animal failed to detect the stimulus on two out of three successive occasions; the intensity was then increased and the procedure repeated. Threshold was defined as that intensity lying midway between the lowest stimulus value correctly reported and the value directly below that. The results were very similar for the different species (with just an indication of slightly reduced sensitivity for the squirrel monkey at the lower frequencies) and did not seem to depend on the behavioral training method used. Stimulation in the experiment was free-field and acoustic measurement (not described in the published article) was done in the same manner as Seiden (averaged sound pressure from different cage positions).

In 1966, Stebbins *et al.* presented data for cynomolgus and pigtailed *(Macaca nemestrina)* macaques in a study designed to determine the upper frequency limit of hearing for this genus. Our animals were able to respond to pure tones as high as 45 kHz but with considerably diminished sensitivity (see Fig. 3). Failure to respond to higher frequencies led us to conclude that 45 kHz was a reasonable estimate of the macaques' upper limit. Our animals were young adolescents, however, and it is possible that younger animals are able to hear tones slightly higher in frequency. The threshold data for the lower frequencies are not only in good agreement with those of Elder and Wendt, but also with data from the recent experiments described above.

Our animals were fitted with earphones which were calibrated at each test frequency on the monkey with a condenser microphone and wave analyzer. The animals learned to press one key, and this response was followed intermittently by a 3-sec tone presentation. One response on a second key while the tone was on produced food. Early or late responses to the tone (e.g., false positive responses) on this second key were punished by a 3–10-sec time-out from the experiment. Guess rate, the frequency with which the animal switched to the second key independently of the tone, was assessed by occasionally presenting "catch" trials on which no tone was presented. The testing method taken from human psychophysics was that of constant stimuli. Several preselected values of stimulation (bracketing the estimated threshold) were presented, and the stimulus intensity to which the animal responded 50% of the time (on the second key) defined the threshold.

Very recently, Heffner *et al.* (1969a, b) have reported complete audiometric functions for two species of prosimians—the tree shrew *(Tupaia glis)* and the galago or bush-baby *(Galago senegalensis).* Most interesting is the fact that these prosimians were responsive to auditory frequencies above 60 kHz—or in the range of some of the lower mammals such as the cat and rat. Maximum sensitivity for the prosimians centers between 8 and 16 kHz and they are somewhat less sensitive than the Anthropoidea (about 10–20 dB) at the lower frequencies.

Sound stimuli were presented free-field and measured with a condenser microphone in the position occupied by the animal's head when it was engaged in making the reporting response. This response consisted of licking a tube, and thus the animal's head was in a relatively fixed position relative to the sound source on successive trials. Behavior was maintained on a variable-ratio schedule of food reinforcement. A conditioned suppression procedure (Hendricks, 1966; Sidman, Ray, Sidman, & Klinger, 1966; Masterton *et al.*, 1969) was then added, in which a tone served as the conditioned suppressor. Threshold testing was carried out with the method of constant stimuli. Threshold was defined as a 50% decrease in the ratio of the number of responses when the tone was present to the number in an equal time period immediately preceding the tone.

A report of some preliminary findings on two subspecies of lemur (Mitchell, Vernon, & Herman, 1970) indicates that their hearing can be considered similar to that of the other prosimians with the exception of their reduced sensitivity around 8 kHz. The behavioral procedure was similar to ours for the macaque (Stebbins *et al.,* 1966) with the important difference that in their experiment electric shock was administered to the animal for incorrectly reporting a tone.

Responding on one bar turned on the tone; one response on a second bar during the tone resulted in food reward. Failure to respond during the tone simply terminated the trial.

Although the laboratory evidence for auditory sensitivity in nonhuman primates is far from complete, a clear picture is beginning to emerge. Many species remain to be tested; additional tests of the generality of existing results with different procedures in different laboratories are necessary. With respect to primate evolution from the Tupaiiformes to man, it would appear that the slight increase in absolute sensitivity of man below 8 kHz is more than offset by his inferiority as a receiver of high-frequency stimulation. It is encouraging to find relatively good agreement (for example, for the macaque) among experimental findings where procedures have been so different.

B. Differential Sensitivity

Perhaps the earliest laboratory experiment on hearing in nonhuman primates was concerned with differential sensitivity to intensity and frequency of acoustic stimulation and was carried out by Shepherd (1910) in the Psychological Laboratory at George Washington University. Because of its charm, precision, and ingenuity, part of his procedure is reported here.

> The apparatus used (for intensity discrimination) was a wooden box 22 x 18 x 10 inches and a small board or slat 18 x 3½ x ⅝ inches arranged to strike the box and thus make a noise. One end of the board or slat was fastened to the top of the box by a leather hinge. By raising the free end of the slat and suddenly letting it go, it struck the top of the box and made a sound varying in loudness with the force with which it struck. To give two sounds of different degrees of intensity or loudness two small sticks, one 3 inches in length, the other 5 inches in length, were separately used to be placed perpendicular to the box and under the free end of the board. By pressing slightly on the slat near the hinge, and suddenly removing the shorter stick, the board would strike the box and produce a noise of a noticeable intensity, and by pressing on the board as before, and withdrawing the larger stick that had been placed at the free end, the board would strike the box and produce a much louder noise [p. 26].

For frequency discrimination,

> An ordinary German mouth harp or harmonica A was used. When I sounded the higher note A_3, the monkey was to go upon the platform used in the preceding noise tests, and was fed there when the note was sounded. When the lowest note, A (two octaves lower), was sounded, he was not to go up and was not fed. The notes were sounded in an irregular order so the animal might not react in a rhythm to the sounds. Care was taken to sound the notes with the same degree of intensity, as nearly as possible. I took the usual precautions that the animal should not obtain a cue from my looks, motions or in any other manner, and react to these stimuli rather than the tones [p. 28].

Shepherd (1910) tells us that his monkeys, the popular rhesus monkey again, were able to learn to discriminate "quantitative differences in noises" and "musical notes of widely different pitch with considerably more facility than do raccoons in similar tests [p. 29]."

Fortunately, we can offer more recent evidence for differential sensitivity to auditory frequency and intensity in nonhuman primates. There are, of course, numerous experiments on auditory frequency discrimination in the monkey (Wegener, 1964). Discrimination between two (often widely separated) frequencies has been a popular test of brain function with the lesion method. Such studies have yielded little useful evidence on differential acuity of nonhuman primates, and for that reason will not be discussed here.

Vernon (1967) has suggested that the most interesting information about hearing of nonhuman primates will come from studies that have determined their frequency difference thresholds particularly at the higher frequencies. The question arises as to whether the frequencies in the ultrasonic range sound alike to the monkey, or if it can distinguish between them. The answer has recently been provided by Heffner et al. (1969a, b), and in our laboratory (Stebbins, Pearson, & Moody, 1970). The evidence clearly indicates that nonhuman primates have very respectable differential acuity at the upper extent of their hearing range. Their Weber ratios ($\Delta f/f$) are well below 0.05 at the very highest frequencies. At 30 kHz, the macaque can differentiate between tones that are less than 500 Hz apart. Heffner's data for the tree shrew and galago (Heffner et al., 1969a, b), together with our data for the macaque, and recent experimental data for man (Filling, 1958), are presented in Fig. 4. The behavioral training and psychophysical testing procedures used by both investigators were very similar to those used in the respective absolute-threshold testing situations. In the conditioned suppression procedure used by Heffner and colleagues, the base or standard frequency was a pulsed tone and was not paired with shock when presented alone. The conditioned stimulus for shock was a compound stimulus with the same standard frequency tone alternating with a comparison tone in a 10-sec pulse train. The frequency of the comparison tone was changed on successive presentations using the method of constant stimuli. The difference threshold for frequency (Δf) was based on a change in response rate during the conditioned stimulus and calculated in the same manner as the absolute thresholds under the suppression procedure described previously. In our laboratory, monkeys were trained to make an observing response that was followed by a change in the frequency of a continuously present, pulsed pure tone. The stimulus change consisted of two alternations of the background or base tone with a tone of different frequency (the comparison tone). A correct report by the animal of the alternation of this standard tone with the comparison tone was reinforced with food. Failure to report had no consequences but an incorrect report produced a brief time-out from the

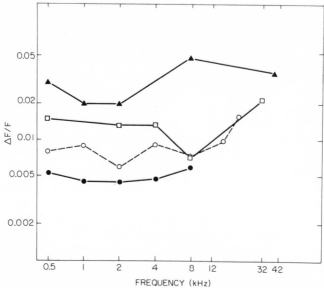

FIG. 4. Frequency discrimination functions at 40 dB above threshold for the tree shrew [*Tupaia* (▲)] (Heffner *et al.*, 1969a), galago [*Galago* (□)] (Heffner *et al.*, 1969b), macaque [*Macaca* (●- - -●)] (Stebbins *et al.*, 1970), and man [*Homo* (●—●)] (Filling, 1958).

experiment. The frequency of the comparison tone was changed on successive trials and difference thresholds were determined by both tracking and constant stimulus methods. It is worth noting that although the behavioral training methods of Heffner *et al.* (1969a, 1969b, 1970) were quite different than our own, the psychophysical procedure and the mode of stimulus presentation were essentially the same. The discriminability functions were obtained at equal sound-pressure levels above threshold (40 dB at each frequency), so that any discriminations that the subjects made could be ascribed to frequency disparity and not differences in loudness. The necessary assumption is that the functions for equal loudness in nonhuman primates parallel the absolute threshold functions as they do in man.

If it is true that nonhuman primates not only are capable of hearing high frequencies, but, with some reasonable degree of precision, can discriminate between them, then this raises some interesting questions concerning ecological significance for the primatologist. Other mammals use very high-frequency information for echolocation. Is it possible that some of the nocturnal or crepuscular prosimians, for example, are either able to echolocate or in some way make use of high-frequency information in transporting themselves around their environment?

Furthermore we have been able to show that the pigtailed macaque can discriminate intensity differences for pure tones over a wide range of frequencies

(Stebbins *et al.,* 1970) with an acuity close to that of man. The procedure was identical to that used in the determination of the frequency difference thresholds; the standard and comparison differed in sound pressure rather than frequency. The difference thresholds for intensity at 40 and 60 dB above absolute threshold for the two macaques tested and corresponding data for man (Riesz, 1928) are shown in Fig. 5. The perturbations in the functions for each

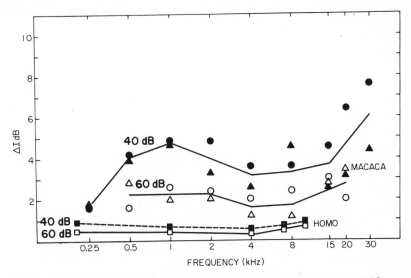

FIG. 5. Intensity discrimination functions at 40 dB and 60 dB above threshold for monkey *(Macaca)* (Stebbins *et al.,* 1970) and man *(Homo)* (Riesz, 1928).

monkey are reliable, and probably reflect individual differences that are not visible in the averaged functions for man. We have fitted functions to the points for the two monkeys by visual inspection as our best estimate of what the characteristic intensity threshold functions for the macaque might look like. The functions are no more than 4 dB above man's at any frequency. Interestingly enough, although the monkey's frequency range of hearing is more extensive than man's and his ability to detect minimum acoustic energy levels (absolute thresholds) from the lowest audible frequencies to about 8 kHz is comparable, his capacity to differentiate sound amplitude and frequency differences is slightly but reliably inferior to that of the average human listener. Confirmation of these findings has been provided by Gourevitch (1970) in some preliminary measurements of critical bands in the monkey. He has pointed out that the human ear appears to be the most selectively tuned with respect to frequency of any animal thus far studied.

Laboratory evidence on other characteristics of the hearing of nonhuman primates is sketchy and incomplete. Earlier, Kuroda (1939) had reported on the ability of female monkeys to localize the call of their infants, and concluded that they could localize sound as well as man. Wegener (1964; in preparation) has compared the sound-localizing capability of man, macaque monkey, and cat. Under one set of experimental conditions, the monkeys were found to be slightly inferior to man and cat, and with limited training, were able to discriminate objects separated by about 8° from the midline or azimuth 75% of the time, whereas humans were able to discriminate them when separated by 2.5° at the same level of accuracy. It is possible that with more extended training, the monkeys might have attained a performance level closer to that of man.

As we have pointed out earlier, loudness, and for that matter, all of the traditional "psychological" or "subjective" attributes of sensation, such as pitch, brightness, and hue, are inextricably tied to spoken language. Consequently, there is no way in which a direct translation of loudness, for example, can be made for purposes of investigating it in nonhuman primates. One must assume first that there is more to loudness than a set of verbal instructions responded to in kind by a human subject, and that loudness has some generality across various species.

In the laboratory, I have attempted to deal with the problem of loudness measurement in the macaque monkey with the use of a simple reaction-time procedure (Stebbins, 1966). I made use of the fact that the latency of a simple, conditioned operant response varies in an orderly fashion with the intensity of the discriminative stimulus for that response. First, response-latency versus stimulus-intensity functions were obtained at several pure-tone frequencies. From these data, equal latency contours were constructed, and I suggested that these might be the nonhuman primate equivalent of equal loudness functions for man (Fletcher & Munson, 1933). Hopefully, further work and additional evidence may clarify the issue and make possible an examination of the way in which animals quantify or scale supraliminal stimulation (see Moody, 1970).

C. Species Comparisons

Recently, Masterton et al. (1969) have attempted to infer the successive stages in the evolution of human hearing on the basis of the behavioral or audiometric findings in mammals. They have suggested that high-frequency hearing is a primitive characteristic, and that it may be the result of selective pressure for accurate sound localization. They have also suggested that low-frequency hearing and absolute sensitivity have developed gradually, with the latter remaining unchanged since its appearance in early members of the Anthropoidea. The final stages of development of human hearing have taken place in the primates, and, thus, it is not surprising that the differences in auditory sensitivity between man and nonhuman primate are small.

It seems safe to assert that there has been a gradual loss in receptivity to high-frequency acoustic stimulation within the primate series. This is perhaps the most striking change in the evolution of the primates, and is clearly seen in Fig. 6 where the threshold functions for five species representing the two suborders are plotted. Some small liberties have been taken with these functions. The data points are not included. The functions for the *Prosimii* (Heffner *et al.,*

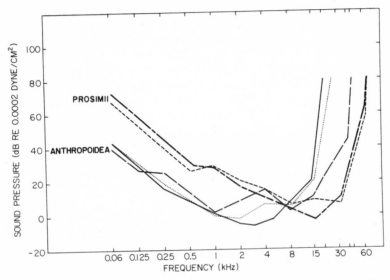

FIG. 6. Auditory threshold functions for the tree shrew [*Tupaia* (– - –)] (Heffner *et al.,* 1969a), galago [*Galago* (- - -)] (Heffner *et al.,* 1969b), macaque [*Macaca* (– – –)] (Stebbins *et al.,* 1966), chimpanzee [*Pan* (· · · · ·)] (Elder, 1934; Farrer and Prim, 1965), and man [*Homo* (——)] (Sivian and White, 1933; Farrer and Prim, 1965).

1969a, b; Masterton *et al.,* 1969) have been extrapolated from 250 to 60 Hz. The curve for the chimpanzee is a combination of Elder's (1934) data at the lower frequencies, and the findings of Farrer and Prim (1965) for the upper-frequency limit. In the same manner, the results for man are from Sivian and White (1933) at the lower frequencies, and Farrer and Prim (1965) at the highest frequency. The data for the macaque are from Stebbins *et al.* (1966).

Although it would appear from Fig. 6 that there is an orderly decrease in the upper-frequency limit of hearing from prosimian to monkey to ape to man, Heffner and Masterton (1970) report that this limit, for at least two prosimian genera, slow lorises *(Nycticebus)* and pottos *(Perodicticus),* may be no higher than that for the macaque (about 40–45 kHz). Perhaps the second most obvious difference (see Fig. 6) is the sensitivity disparity between the two primate suborders at the lower frequencies. Below 1 kHz and extending down to 60 Hz (a range of 4 octaves), the tree shrew and the galago seem to be less sensitive, with a constant 20 dB difference in threshold, than the macaque, chimpanzee, or man.

In addition, Masterton *et al.* (1969) have argued for a decline in the frequency of the point of maximum sensitivity in the evolution of human hearing. This may be difficult to see if only the primates are considered; in fact, there is a clear indication, for example, that the macaque has at least two points of maximum sensitivity at 1 and 8 kHz.

Clearly, this discussion of evolutionary development and species comparisons of hearing has to be considered incomplete. Only three of the six prosimian families have been examined. Among the apes, we have no information on the gibbon (Family: *Hylobatidae*), orangutan *(Pongo)* or gorilla *(Gorilla)*. In addition, it is very likely that some of our conceptions about the patterns of development and the differences between suborders and families discussed here and by Masterton *et al.* (1969) will change as we acquire information about auditory sensitivity from a wider range of nonhuman primates.

V. SUMMARY

It is our contention that we have progressed significantly in our understanding of primate hearing since the publication of the book by Yerkes and Yerkes in 1929, and that we now have valid and reliable experimental evidence on the acoustic sensitivity of nonhuman primates. There are, however, other aspects of hearing besides absolute sensitivity that need to be considered, and there are species and entire families of primates whose hearing is yet unexplored.

Although the object of the chapter has been to present laboratory findings, there has been an attempt to present a sizeable amount of methodology basic to the laboratory work. First, it is helpful in understanding the evidence if one has an appreciation of the methods which have produced it; second, there are some complex and tricky problems in working with acoustic energy and with behavioral conditioning and testing procedures which need to be considered by anyone who would attempt to work, or to evaluate work, in primate hearing.

There are some intriguing biological questions which should receive more attention in the future. Not the least of these is the general relation between structure and function which Pocock (1905) and Sonntag (1924) considered, although naively, almost half a century ago. For example, among the tree shrews, *Ptilocercus* is nocturnal and possesses good control over its fairly extensive ear musculature in contrast to the diurnal *Tupaia* which has a small, relatively immobile pinna. We have no audiometric data on *Ptilocercus,* and it is easy to speculate that their hearing is superior to *Tupaia*—but perhaps only in their ability to localize sound.

With increased activity in the study of communication, and the natural conditions under which it occurs in the primate social group, it should soon be possible to arrive at a better understanding of the relationship between auditory acuity and the specific requirements on hearing demanded by the environment. There are also specific capabilities, such as sound localization and frequency discrimination, which require further study in the laboratory.

Finally, the need for a closer, working relationship is evident between the experimental psychologist (whose work we have described here) and the physiologist and anatomist, who have begun to examine in detail the cochlear end-organ and the central auditory pathways in the primates, and the field primatologist, who is using increasingly sophisticated instrumentation and technique in the field in analyzing primate vocal communication.

REFERENCES

Allen, J. N., Dalton, L. A., Jr., Henton, W., & Taylor, H. L. Behavioral analysis of auditory functioning in primates. Technical Report No. 68-6, 6571st Aeromedical Research Laboratory, Holloman Air Force Base, New Mexico, 1968.

Behar, I., Cronholm, J. N., & Loeb, M. Auditory sensitivity of the rhesus monkey. *Journal of Comparative & Physiological Psychology*, 1965, **59**, 426-428.

Beranek, L. L. *Acoustic measurements.* New York: John Wiley, 1949.

Berkley, M. A. Visual discriminations in the cat. *In* W. C. Stebbins (ed.), *Animal psychophysics: The design and conduct of sensory experiments.* New York: Appleton-Century-Crofts, 1970. Pp. 231-247.

Blough, D. S. The study of animal sensory processes by operant methods. *In* W. K. Honig (ed.), *Operant behavior: Areas of research and application.* New York: Appleton-Century-Crofts, 1966. Pp. 345-379.

Blough, D. S., & Yager, D. Visual psychophysics in animals. *In* L. M. Hurvich and Dorothea Jameson (eds), *Handbook of sensory physiology,* Volume VII, Part 3, *Visual psychophysics.* Berlin: Springer-Verlag, 1971, in press.

Clack, T. D., & Herman, P. N. A single-lever psychophysical adjustment procedure for measuring auditory thresholds in the monkey. *Journal of Auditory Research,* 1963, **3**, 175-183.

Culler, E., Finch, G., Girden, E., & Brogden, W. J. Measurements of acuity by the conditioned-response technique. *Journal of General Psychology*, 1935, **12**, 223-227.

Dalland, J. I. Hearing sensitivity in bats. *Science*, 1965, **150**, 1185-1186.

Dalton, L. W., Jr. Summed evoked cortical responses to pure-tone stimuli in the rhesus *(Macaca mulatta)* monkey. Technical Report No. 67-3, 6571st Aeromedical Research Laboratory, Holloman Air Force Base, New Mexico, 1967.

Dalton, L. W., Jr. Auditory sensitivity in the rhesus *(Macaca mulatta)* and the white-throated capuchin *(Cebus capuchinus)* monkeys: A comparison of three techniques. Technical Report No. 68-14, 6571st Aeromedical Research Laboratory, Holloman Air Force Base, New Mexico, 1968.

Donisthorpe, Jill H. A pilot study of the mountain gorilla *(Gorilla gorilla beringei)* in South West Uganda, February to September, 1957. *South African Journal of Science,* 1958, **54**, 195-217.

Elder, J. H. Audiometric studies with the chimpanzee. *Psychological Bulletin,* 1933, **30,** 547-548.

Elder, J. H. Auditory acuity of the chimpanzee. *Journal of Comparative Psychology,* 1934, **17,** 157-183.

Elder, J. H. The upper limit of hearing in chimpanzee. *American Journal of Physiology,* 1935, **112,** 109-115.

Estes, W. K., & Skinner, B. F. Some quantitative properties of anxiety. *Journal of Experimental Psychology,* 1941, **29,** 390-400.

Farrer, D. N., & Prim, M. M. A preliminary report on auditory frequency threshold comparisons of humans and pre-adolescent chimpanzees. Technical Report No. 65-6, 6571st Aeromedical Research Laboratory, Holloman Air Force Base, New Mexico, 1965.

Filling, S. Studies on a series of normal subjects and on a series of patients from a hearing rehabilitation centre. *Difference limen for frequency.* 1 Odense, Denmark: Andelsbogtrykkeriet, 1958.

Fletcher, H. *Speech and Hearing.* New York: Van Nostrand Co., 1929.

Fletcher, H., & Munson, W. A. Loudness, its definition, measurement, and calculation. *Journal of the Acoustical Society of America,* 1933, **5,** 82-108.

Fujita, S., & Elliott, D. N. Thresholds of audition for three species of monkey. *Journal of the Acoustical Society of America,* 1965, **37,** 139-144.

Gourevitch, G. Detectability of tones in quiet and in noise by rats and monkeys. *In* W. C. Stebbins (ed.), *Animal psychophysics: The design and conduct of sensory experiments.* New York: Appleton-Century-Crofts, 1970. Pp. 67-97.

Gourevitch, G., & Hack, M. N. Audibility in the rat. *Journal of Comparative & Physiological Psychology,* 1966, **62,** 289-291.

Hanson, H. Psychophysical thresholds and aversive control. Paper read at meetings of American Association for the Advancement of Science, Washington, D.C., December, 1966.

Harris, J. D. The auditory acuity of pre-adolescent monkeys. *Journal of Comparative Psychology,* 1943, **35,** 255-265.

Heffner, H. E., & Masterton, B. Hearing in primitive primates: Slow loris *(Nycticebus coucang)* and potto *(Perodicticus potto). Journal of Comparative & Physiological Psychology,* 1970, **71,** 175-182.

Heffner, H., Ravizza, R., & Masterton, B. Hearing in primitive mammals, III: Tree shrew *(Tupaia glis). Journal of Auditory Research,* 1969, **9,** 12-18. (a)

Heffner, H., Ravizza, R., & Masterton, B. Hearing in primitive mammals, IV: Bushbaby *(Galago senegalensis). Journal of Auditory Research,* 1969, **9,** 19-23. (b)

Hendricks, J. Flicker thresholds as determined by a modified conditioned suppression procedure. *Journal of the Experimental Analysis of Behavior,* 1966, **9,** 501-506.

Hirsh, I. J. Audition. *In* J. B. Sidowski (ed.), *Experimental methods and instrumentation in psychology.* New York: McGraw-Hill, 1966. Pp. 247-271.

Kalischer, O. Uber die Tondressur der Affen. *Zentralblatt für Physiologie,* 1912, **26,** 713-714.

Keast, D. N. *Measurements in mechanical dynamics.* New York: McGraw-Hill, 1967.

Kelleher, R. T. Operant conditioning. *In* A. M. Schrier, H. F. Harlow, and F. Stollnitz (eds), *Behavior of nonhuman primates,* Vol. 1. New York: Academic Press, 1965. Pp. 211-247.

Keller, F. S., & Schoenfeld, W. N. *Principles of psychology.* New York: Appleton-Century-Crofts, 1950.

Krasnegor, N. A., & Brady, J. V. The effects of signal frequency and shock probability on a prolonged vigilance task. Paper read at meetings of American Psychological Association, San Francisco, September, 1968.

Kuroda, R. On sound localization in a monkey. *Acta Psychologica Keijo*, 1939, **3**, 74-85.

Lawrence, M., & Yantis, P. A. Onset and growth of aural harmonics in the overloaded ear. *Journal of the Acoustical Society of America*, 1956, **28**, 852-858.

Marler, P. Communication in monkeys and apes. *In* I. DeVore (ed.), *Primate behavior: Field studies of monkeys and apes*. New York: Holt, Rinehart & Winston, 1965. Pp. 514-543.

Martin, P., Romba, J. J., & Gates, H. W. A method for the study of hearing loss and recovery in rhesus monkeys. Laboratory Technical Memorandum 11-62, OCMS Code 5010. 11.841A. U.S. Army Human Enginr., 1962.

Masterton, B., Heffner, H., & Ravizza, R. The evolution of human hearing. *Journal of the Acoustical Society of America*, 1969, **45**, 966-985.

Mitchell, C., Gillette, R., Vernon, J., & Herman, P. Pure-tone auditory behavioral thresholds in three species of lemurs. *Journal of the Acoustical Society of America*, 1970, **48**, 531-535.

Moody, D. B. Reaction time as an index of sensory function. *In* W. C. Stebbins (ed.), *Animal psychophysics: The design and conduct of sensory experiments*. New York: Appleton-Century-Crofts, 1970. Pp. 277-302.

Nevin, J. A. A method for the determination of psychophysical functions in the rat. *Journal of the Experimental Analysis of Behavior*, 1964, **7**, 169.

Pavlov, I. P. *Conditioned reflexes. An investigation of the physiological activity of the cerebral cortex.* (G. B. Anrep, transl. and Ed.) London: Oxford Univ. Press, 1927.

Pocock, R. I. Observations upon a female specimen of the Hainan gibbon *(Hylobates hainanus)* now living in the Society's gardens. *Proceedings of the Zoological Society of London*, 1905, **2**, 169-180.

Ray, B. A. Psychophysical testing of neurological mutant mice. *In* W. C. Stebbins (ed.), *Animal psychophysics: The design and conduct of sensory experiments.* New York: Appleton-Century-Crofts, 1970. Pp. 99-124.

Riesz, R. R. Differential intensity sensitivity of the ear for pure tones. *Physiological Reviews*, 1928, **31**, 867-875.

Ross, S. Matching functions and equal-sensation contours for loudness. *Journal of the Acoustical Society of America*, 1967, **42**, 778-793.

Schaller, G. B. *The mountain gorilla: Ecology and behavior.* Chicago, Illinois: Univ. Chicago Press, 1963.

Seiden, H. R. *Auditory acuity of the marmoset monkey* (Hapale jacchus). (Doctoral dissertation, Princeton University.) Ann Arbor, Mich.: University Microfilms, 1958, No. 58-7888.

Semenoff, W. A., & Young, F. A. Comparison of the auditory acuity of man and monkey. *Journal of Comparative & Physiological Psychology*, 1964, **57**, 89-93.

Shepherd, W. T. Some mental processes of the rhesus monkey. *Psychological Review Monographs*, 1910, **12**, No. 52.

Sidman, M., Ray, B. A., Sidman, R. L., & Klinger, J. M. Hearing and vision in neurological mutant mice: A method for their evaluation. *Experimental Neurology*, 1966, **16**, 377-402.

Sivian, L. J., & White, S. D. On minimum audible sound fields. *Journal of the Acoustical Society of America*, 1933, **4**, 288-321.

Skinner, B. F. *The behavior of organisms.* New York: Appleton-Century-Crofts, 1938.

Smith, J. C., & Tucker, D. Olfactory mediation of immediate X-ray detection. *In* C. Pfaffmann (ed.), *Olfaction and taste*, III. New York: Rockefeller Univ. Press, 1969.

Sonntag, C. F. *The morphology and evolution of apes and man.* London: John Bale, Sons & Danielsson Ltd., 1924.

Stebbins, W. C. Auditory reaction time and the derivation of equal loudness contours for the monkey. *Journal of the Experimental Analysis of Behavior*, 1966, **7**, 135-142.

Stebbins, W. C. (ed.) *Animal psychophysics: The design and conduct of sensory experiments.* New York: Appleton-Century-Crofts, 1970. (a)

Stebbins, W. C. Studies of hearing and hearing loss in the monkey. *In* W. C. Stebbins (ed.), *Animal psychophysics: The design and conduct of sensory experiments.* New York: Appleton-Century-Crofts, 1970. Pp. 41-66. (b)

Stebbins, W. C., Green, S., & Miller, F. L. Auditory sensitivity of the monkey. *Science,* 1966, **153,** 1646-1647.

Stebbins, W. C., Pearson, R. D., & Moody, D. B. Hearing in the monkey *(Macaca)*: absolute and differential sensitivity. *Journal of the Acoustical Society of America,* 1970, **47,** 67. (Abstract)

Stebbins, W. C., Miller, J. M., Johnsson, L-G., & Hawkins, J. E. Ototoxic hearing loss and cochlear pathology in the monkey. *Annals of Otology, Rhinology, and Laryngology,* 1969, **78,** 1007-1023.

Struhsaker, T. T. Auditory communication among vervet monkeys *(Cercopithecus aethiops). In* S. A. Altmann (ed.), *Social communication among primates.* Chicago: Univ. Chicago Press, 1967. Pp. 281-324.

Struhsaker, T. T. Notes on the behavioral ecology of *Galagoides demidovii,* in Cameroon, West Africa. *Mammalia,* 1970, **34,** 207-211.

Terrace, H. S. Stimulus control. *In* W. K. Honig (ed.), *Operant behavior: Areas of research and application.* New York: Appleton-Century-Crofts, 1966. Pp. 271-344.

Vernon, J. Hearing in subhuman primates. *Primate News,* 1967, **5,** 4-10.

Wallis, H. M. On the growth of hair upon the human ear, and its testimony to the shape, size, and position of the ancestral organ. *Proceedings of the Zoological Society of London,* 1897, 298-310.

Warwick, J. E. The habits and manners of the female Borneo orangutan *(Simia satyrus)* and the male chimpanzee *(Simia troglodytes)* as observed during their exhibition at the Egyptian Hall in 1831. *Magazine of Natural History,* 1832, **5,** 305-309.

Wegener, J. G. Auditory discrimination behavior of normal monkeys. *Journal of Auditory Research,* 1964, **4,** 81-106.

Wendt, G. R. Auditory acuity of monkeys. *Comparative Psychology Monographs,* 1934, **10,** No. 4.

Woodworth, R. S., & Schlosberg, H. (eds), *Experimental psychology (Rev. ed.).* New York: Holt & Co., 1954.

Yager, D. Behavioral measures of the spectral sensitivity of the dark-adapted goldfish. *Nature,* 1968, **220,** 1052-1053.

Yerkes, R. M., & Yerkes, A. W. *The great apes.* New Haven, Conn.: Yale Univ. Press, 1929.

Author Index

Numbers in italics refer to the pages on which the complete references are listed.

Subject Index

SUMMARY OF PRIMATE CLASSIFICATION

Suborder	Superfamily	Family	Subfamily	Genus	Vernacular name(s)
Prosimiae (prosimians)	Tupaioidea	Tupaiidae	Tupaiinae	*Tupaia* *Dendrogale* *Urogale*	Tree shrew Smooth-tailed tree shrew Philippine tree shrew
			Ptilocercinae	*Ptilocercus*	Pen-tailed tree shrew
	Lemuroidea	Lemuridae	Lemurinae (greater lemurs)	*Lemur* *Hapalemur* *Lepilemur*	Lemur Gentle lemur Sportive lemur
			Cheirogaleinae (lesser lemurs)	*Cheirogaleus* *Microcebus*	Dwarf lemur, mouse lemur
		Indridae	Indriinae	*Avahi* *Propithecus* *Indri*	Woolly lemur Sifaka Indri
		Daubentoniidae		*Daubentonia*	Aye-aye
	Lorisoidea	Lorisidae		*Loris* *Nycticebus* *Arctocebus* *Perodicticus*	Slender loris Slow loris Angwantibo Potto
		Galagidae		*Galago*	Galago (bush-baby)
	Tarsioidea	Tarsiidae		*Tarsius*	Tarsier
		Callithricidae		*Callithrix* *Leontocebus*	Marmoset Tamarin, pinche
			Callimiconinae	*Callimico*	Goeldi's "marmoset"
				Aotes	Douroucouli (night monkey, owl monkey)
				Callicebus	Titi